Feedback Control Systems

The MATLAB®/Simulink® Approach

Synthesis Lectures on Control and Mechatronics

Editors
Chaouki Abdallah, *Georgia Institute of Technology*
Mark Spong, *University of Texas at Dallas*

Feedback Control Systems: The MATLAB®/Simulink® Approach

Farzin Asadi, Robert E. Bolanos, and Jorge Rodríguez

ISBN: 978-3-031-00703-3   paperback
ISBN: 978-3-031-01831-2   ebook
ISBN: 978-3-031-00084-3   hardcover

DOI 10.1007/978-3-031-01831-2

A Publication in the Springer series
SYNTHESIS LECTURES ON CONTROL AND MECHATRONICS

Lecture #5
Series Editors: Chaouki Abdallah, *Georgia Institute of Technology*
                Mark Spong, *University of Texas at Dallas*
Series ISSN
Print 1939-0564   Electronic 1939-0572

Feedback Control Systems

The MATLAB®/Simulink® Approach

Farzin Asadi
Kocaeli University, Kocaeli, Turkey

Robert E. Bolanos
Southwest Research Institute, Texas, U.S.A.

Jorge Rodríguez
Power Smart Control, Leganés, Spain

SYNTHESIS LECTURES ON CONTROL AND MECHATRONICS #5

ABSTRACT

Feedback control systems is an important course in aerospace engineering, chemical engineering, electrical engineering, mechanical engineering, and mechatronics engineering, to name just a few. Feedback control systems improve the system's behavior so the desired response can be acheived.

The first course on control engineering deals with Continuous Time (CT) Linear Time Invariant (LTI) systems. Plenty of good textbooks on the subject are available on the market, so there is no need to add one more. This book does not focus on the control engineering theories as it is assumed that the reader is familiar with them, i.e., took/takes a course on control engineering, and now wants to learn the applications of MATLAB® in control engineering. The focus of this book is control engineering applications of MATLAB® for a first course on control engineering.

KEYWORDS

closed-loop control, control, controller, feeback, PID controller, MATLAB®, Simulink®

Contents

Preface

This book is composed of five chapters. Here is a brief summary of these chapters:

Chapter 1: Introduction to MATLAB®. This chapter introduces the MATLAB® for novices. The reader learns the fundamental concepts of MATLAB®.

Chapter 2: Commonly Used Commands in Analysis of Control Systems. This chapter studies some of the most important commands of Control Systems Toolbox™. Commands such `tf, zpk, ss, bode, nyquist, rlocus, impulse, step, feedback, seriesi, parallel`, etc. are studied in this chapter.

Chapter 3: Introduction to Simulink®. Simulink® is one of the most important tools for simulation of dynamical systems. Simulink® is a graphical environment. The user makes the model of system using plenty of ready to use blocks. So, thanks to the powerful library of blocks Simulink® has, the user can simulate even complex systems easily without dealing with coding. This chapter introduces the Simulink® with some examples.

Chapter 4: Controller Design in MATLAB®. Computers are an important part of modern engineering sciences. Nearly all the engineering designs are done with the aid of computers. Control engineering is not an exception at all. MATLAB's Control System Toolbox provides plenty of functions to the design of control systems.

About 90% of controllers in industry are PID (Proportional-Integral-Derivative). Keeping this in mind, the chapter starts with pidTuner command, a powerful command to design PID controllers. The chapter finishes with the Control System Designer app.

Chapter 5: Introduction to System Identification Toolbox™. System Identification makes mathematical models for dynamical systems based on their input/output signals. An illustrative example is studied in this chapter.

Farzin Asadi, Robert E. Bolanos, and Jorge Rodríguez
April 2019

Acknowledgments

The authors gratefully acknowledge MathWorks® support for this project.

Farzin Asadi, Robert E. Bolanos, and Jorge Rodríguez
April 2019

CHAPTER 1

Introduction to MATLAB®

1.1 INTRODUCTION

MATLAB® (MATrix LABoratory) is a multi-paradigm numerical computing environment and proprietary programming language developed by MathWorks. MATLAB® allows matrix manipulations, plotting of functions and data, implementation of algorithms, creation of user interfaces, and interfacing with programs written in other languages, including C, C++, C#, Java, Fortran, and Python.

As of 2018, MATLAB® has more than 3 million users worldwide. MATLAB® users have varied backgrounds in engineering, science, and economics. This chapter is a general introduction to MATLAB®.

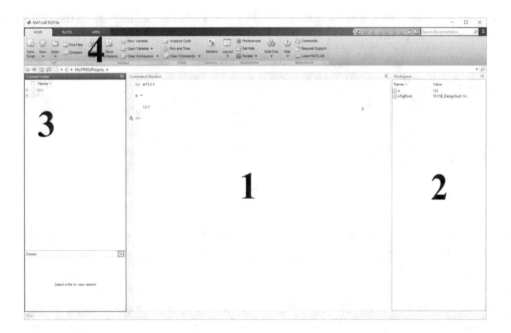

Figure 1.1: Different parts of MATLAB®.

1.2   DIFFERENT PARTS OF MATLAB®

The MATLAB® environment is shown in Fig. 1.1.

The main window of MATLAB® is composed of four parts.

Command Window

Command window is shown with number 1 in Fig. 1.1. The MATLAB® commands are written in this window. After the command is written, the Enter key of the keyboard must be pressed in order to run it (see Fig. 1.2).

Figure 1.2: Command window.

Work Space

Workspace window is shown with number 2 in Fig. 1.1. Variables defined inside the MATLAB® environment are listed in Fig. 1.3.

Workspace	
Name ▲	Value
a	123
b	456

Figure 1.3: Workspace.

Current Folder

Current folder window is shown with number 3 in Fig. 1.1. It shows the current working path. You can set your desired path by clicking the "Browse for folder" icon (see Fig. 1.4).

Menu Bar

Menu bar is shown with number 4 in Fig. 1.1. It provides easy access to some of the widely used operations (see Fig. 1.5). For instance, in order to enter the Simulink environment, the user must type the word "simulink" in the command window. The user can get rid of writing the word simulink by clicking the Simulink icon (Fig. 1.6) in the menu bar.

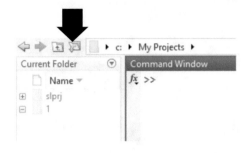

Figure 1.4: Current folder.

Figure 1.5: Menu bar.

Figure 1.6: Simulink icon.

You can obtain the default layout, i.e., layout shown in Fig. 1.7, by clicking the "Default."

1.3   MATLAB'S EDITOR

MATLAB® has a built-in editor which can be used for writing MATLAB® codes. In order to run the editor, write the following command inside the command window:

```
>>edit
```

The editor is shown in Fig. 1.8.

You can save what you wrote inside the editor by using the "Save" icon (Fig. 1.9).

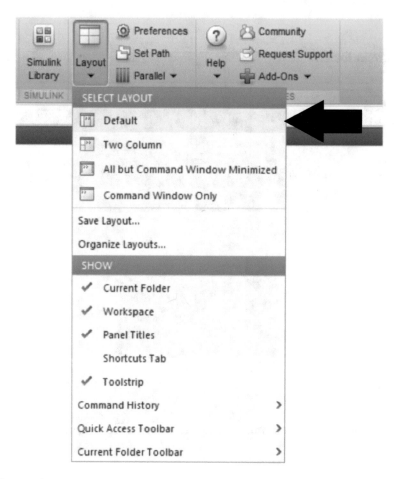

Figure 1.7: Layout icon.

Figure 1.8: MATLAB's editor.

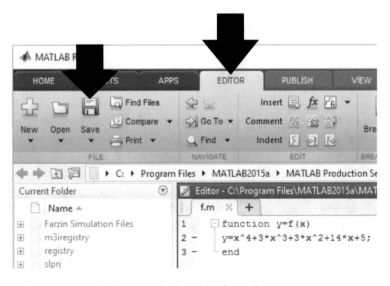

Figure 1.9: Save the content of editor with the aid of save button.

Press the F5 key on the keyboard or click the Run icon in order to run the commands written inside the editor (Fig. 1.10).

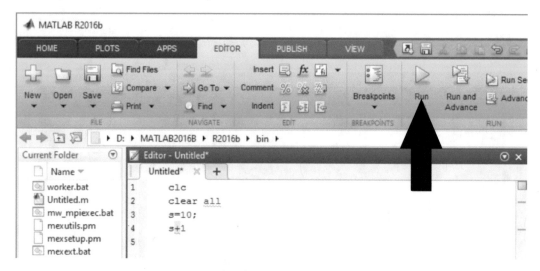

Figure 1.10: Run icon.

1.4   MATLAB'S HELP SYSTEM

You can press the F1 key on your keyboard in order to call the help window shown in Fig. 1.11.

Write the desired keywords in the "Search Documentation" box and press Enter key on your keyboard.

MATLAB® has a command named `help`. For instance, you can see the help for `dir` command by typing `help dir` inside the command window (Fig. 1.12).

You can use https://www.mathworks.com/help/ to get help as well. Write your keywords inside the "Search Rxxxxx Documentation" box (Fig. 1.13).

1.5   MATLAB'S KEYWORDS

MATLAB® has 17 keywords, as shown in Table 1.1.

Keywords are shown in blue either in command window or MATLAB® editor (Fig. 1.14).

Figure 1.11: Help window.

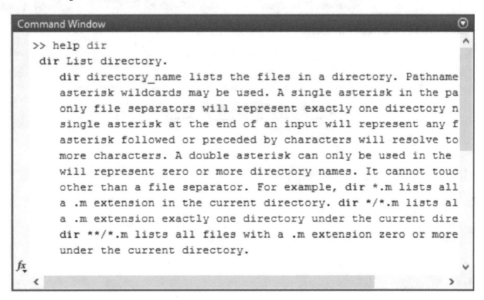

Figure 1.12: You can use the help command to obtain information about a command.

Figure 1.13: MATLAB's online help.

Figure 1.14: Keywords are shown in blue.

Table 1.1: MATLAB® keywords

break	case	catch	continue
else	elseif	end	for
function	global	if	otherwise
persistent	return	switch	try
while			

1.6 MATLAB'S TOOLBOXES

Toolboxes are a set of functions designed for a related purposes. They are sold as a package. Some of the available toolboxes are shown in Table 1.2.

1.7 VARIABLES

Variable names are composed of English characters, numbers, or underline. The variable name must start with an English character. For instance, "_myAge" or "1a" are NOT acceptable.

You can use the isvarname command in order to see if the selected name is acceptable or NOT. For instance, if you write isvarname kp and press the Enter key, the MATLAB® shows 1. This means that the selected name (in this example "kp") is acceptable. When MATLAB® shows 0, it means that the name is NOT acceptable.

MATLAB® is a case-sensitive program, i.e., it is sensitive to small and capital letters. For instance, "my_variable" and "My_variable" is NOT the same.

Some of the MATLAB's simple commands are shown in Table 1.3.

MATLAB® does NOT show the command result on the screen if you put a semicolon at the end of commands.

```
>>Dataspeed=115200;

>>Dataspeed=115200

    Dataspeed =

        115200
```

In order to see the variables defined variables, use the who command, i.e., write who in the command window and press Enter key. If you want to see the size and number of bytes used for variables, use the command whos.

Table 1.2: **Some of MATLAB's toolboxes**

MATLAB	Fixed-Point Designer	Paralle Computing Toolbox
Simulink	Fuzzy Logic Toolbox	Phased Array Toolbox
Aerospace Blockset	Global Optimization Toolbox	Polyphase Bug Finder
Aerospace Toolbox	HDL Coder	Polyphase Code Prover
Anntenna Toolbox	HDL Verifier	Polyphase products for Ada
Audio System Toolbox	IEC Certified Kit	Power Train Blockset
Automated Driving System Toolbox	Image Processing Toolbox	RF Blockset
Bioinformatics Toolbox	Instalation, Licensing and Activation	RF Toolbox
Communications System Toolbox	Instrument Control Toolbox	Risk Management Toolbox
Computer Vision Toolbox	LTE System Toolbox	Robotic Toolbox
Control System Toolbox	Mapping Toolbox	Robust Control Toolbox
Curve Fitting Toolbox	MATLAB Coder	Signal Processing Toolbox
Data Acquisition Toolbox	MATLAB Compiler	SimBiology
Database Toolbox	MATLAB Compiler SDK	SimEvents
Datafeed Toolbox	MATLAB Distributed Computing Toolbox	Simscape
DO Qualification Kit (for DO-178)	MATLAB Production Server	Simscape Driveline
DSP System Toolbox	MATLAB Report Generator	Simscape Electronics
Econometrics Toolbox	Predictive Control Toolbox	Simscape Fluids
Embeded Coder	Model Base calibration Toolbox	Simscape Multibody
Filter Design HDL Coder	Nueral Network Toolbox	Simscape Multibody Link
Financial Instruments Toolbox	OPC Toolbox	Simscape Power Systems
Financial Toolbox	Optimization Toolbox	Simulink 3D Animation
Simulink 3D Animation	Simulink PLC Coder	Symbolic Math Toolbox
Simulink Code Inspector	Simulink Real Time	System Identification Toolbox
Simulink Coder	Simulink Report Generator	Trading Toolbox
Simulink Control Design	Simulink Test	Vehicle Network Toolbox
Simulink Design Optimization	Simulink Verification and Validation	Wavelet Toolbox
Simulink Design Verifier	Spreadsheet link	WLAN System Toolbox
Simulink Desktop Real-Time	Stateflow	

Table 1.3: Some of MATLAB's simple commands

Commands	Explanation
`>>x=2018`	This command asked MATLAB to make a variable named "x". x is 1×1 matrix and contains value 2018.
`>>x`	This command asks MATLAB to show the content of variable "x".
`>>x=x+1`	This command asks MATLAB to increase the value of variable x by 1.
`>>Name= 'John'`	This command asks MATLAB to make a variable named "Name". "Name" is 1×4 matrix contains "John" word.

You can clear the screen by using the command `clc`. In order to clear all the variables, use the command `clear all`. If you want to clear a specific variable use the command `clear`. You must write the variable name after the `clear`.

1.8   BASIC OPERATORS

MATLAB's basic operators are shown in Table 1.4.

Table 1.4: MATLAB's basic operators

Operator	Explanation	Operator	Explanation
+	Addition	^	Matrix power
−	Subtraction	.*	Elementwise multiplication
*	Matrix Multiplication	./	Elementwise right division
/	Division	.\	Elementwise left division
.	Decimal point	.^	Elementwise power
\	Matrix left division	.'	Transpose
=	Assignment		

Example 1.1

```
>>A=[1 2 3];

>>B=[4 5 6];

>>A./B
```

```
ans =

        0.2500       0.4000       0.5000

>>A.^B

ans =

        1       32       729

>>A.*B

ans =

        4       10       18
```

1.9   LOGICAL OPERATORS

MATLAB's logical operators are shown in Table 1.5.

1.10   TRIGONOMETRICS FUNCTIONS

MATLAB's trigonometric functions are shown in Table 1.6.

Example 1.2
Calculate $\sin(x)^4 - 3\cos(x)$ for $x = 30°$.

Solution:
First way:

```
>>x=30*pi/180;

>>sin(x)^4-3*cos(x)

ans=

        -2.5356
```

Table 1.5: MATLAB's logical operators

Operator	Explanation	Example

== or eq	When left and right operand are equal, it returns 1. Otherwise it returns 0.	>> 5==6
		ans =
		    0
		>> 5==5
		ans
		    1

~= or ne	When left and right operand are NOT equal, it returns 1. Otherwise it returns 0.	>> 5~=6
		ans =
		    1
		>> 5~=5
		ans =
		    0

> or gt (greater than)
>= or ge (greater than or equal)
< or lt (less than)
<= or le (less than or equal)
	When the inequality is satisfied, it returns 1. Otherwise it returns 0.	>> 5>=6
		ans =
		    0
		>> 5<=15
		ans =
		    1

& or and	When the two operand are true, it returns 1. Otherwise it returns 0.	>>and(eq(3,3),5>2) and = 1
| or or	When at least one of the two operand are true, it returns 1. Otherwise it returns 0.	>>eq(5,5)|3>12 ans = 1

~ or NOT	Returns the complement of the logical expression.	>>~eq(5,5)
		ans =
		    0
		>>NOT(eq(5,5))
		ans =
		    0

Table 1.6: MATLAB's trigonometric functions (*Continues*)

Function	Explanation	Example

sin	Calculates the sine of an angle. Operand must be given in Radyans.	`>>sin(1)`
		`ans =`
		`      0.8415`

sind	Calculates the sine of an angle. Operand must be given in Degrees.	`>>sind(30)`
		`ans =`
		`      0.5000`

asin	Calculates the arc sine. Result is in Radyans.	`>>asin(0.5)`
		`ans =`
		`      0.5236`

asind	Calculates the arc sine. Result is in Degrees.	`>>asind(.5)`
		`ans =`
		`      30.0000`

cos	Calculates the cosine of an angle. Operand must be given in Radyans.	`>>cos(1)`
		`ans =`
		`      0.5403`

cosd	Calculates the cosine of an angle. Operand must be given in Degrees.	`>>cosd(.5)`
		`ans =`
		`      1.0000`

acos	Calculates the arc cosine. Result is in Radyans.	`>>acos(.5)`
		`ans =`
		`      1.0472`

acosd	Calculates the arc cosine. Result is in Degrees.	`>>acosd(.5)`
		`ans =`
		`      60.0000`

tan	Calculates the tangent of an angle. Operand must be given in Radyans.	`>>tan(1)`
		`ans =`
		`      1.5574`

tand	Calculates the tangent of an angle. Operand must be given in Degrees.	`>>tand(60)`
		`ans =`
		`      1.7321`

Table 1.6: (*Continued*) MATLAB's trigonometric functions (*Continues*)

atan	Calculates the arc tangent. Result is in Radyans.	>>atan(1) ans = 0.7854
atand	Calculates the arc tangent. Result is in Degrees.	>>atand(30) ans = 88.0908
sec	Calculates the secant of an angle. Operand must be given in Radyans.	>>sec(1) ans = 1.8508
secd	Calculates the secant of an angle. Operand must be given in Degrees.	>>secd(10) ans = 1.0154
asec	Calculates the arc secant. Result is in Radyans.	>>asec(30) ans = 1.5375
asecd	Calculates the arc secant. Result is in Degrees.	>>asecd(30) ans = 88.0898
csc	Calculates the cosecant of an angle. Operand must be given in Radyans.	>>csc(1) ans = 1.1884
cscd	Calculates the cosecant of an angle. Operand must be given in Degrees.	>>cscd(1) ans = 57.2987
acsc	Calculates the arc cosecant. Result is in Radyans.	>>acsc(1) ans = 1.5708
acscd	Calculates the arc cosecant. Result is in Degrees.	>>acscd(1) ans = 90
cot	Calculates the cotangent of an angle. Operand must be given in Radyans.	>>cot(1) ans = 0.6421

Table 1.6: (*Continued*) MATLAB's trigonometric functions

cotd	Calculates the cotangent of an angle. Operand must be given in Degrees.	```>>cotd(1)```
		```ans =```
		      ```57.2900```

acot	Calculates the arc cotangent. Result is in Radyans.	```>>acot(1)```
		```ans =```
		      ```0.7854```

acotd	Calculates the arc cotangent. Result is in Degrees.	```>>acotd(60)```
		```ans =```
		      ```0.9548```

Second way:

```
>>x=30;

>>sind(x)^4-3*cosd(x)

ans=

    -2.5356
```

1.11 LOGARITHMIC AND EXPONENTIAL FUNCTIONS

MATLAB's logarithmic and exponential functions are shown in Table 1.7.
NOTE: In MATLAB®, $1.234e5$ equals 1.234×10^5. See Table 1.8 for more examples.

1.12 COMPLEX NUMBERS FUNCTIONS

MATLAB's functions related to complex numbers are shown in Table 1.9.

1.13 ROUNDING FUNCTIONS

MATLAB's rounding functions are shown in Table 1.10.

1.14 REMAINDER FUNCTIONS

MATLAB's remainder functions are shown in Table 1.11.

Table 1.7: MATLAB's logarithmic and exponential functions

Function	Explanation	Example
exp	Calculates e^x. (Note: $e = 2.7182$)	>>exp(2) ans = 7.3891
log	Calculates the $\log_e(x)$	>>log(2.7182) ans = 1.0000
log10	Calculates the $\log_{10}(x)$	>>log10(100) ans = 2
log2	Calculates the $\log_2(x)$	>>log2(4) ans = 2
sqrt	Calculates the $\sqrt{x}$.	>>sqrt(4) ans = 2
power	Calculates a^b, i.e., power (a,b) = a^b.	>>power(2,3) ans = 8

Table 1.8: Conversion of simple expression to MATLAB®

Mathematics	MATLAB
12^{34}	>>12^34
$\left(\frac{1}{2}\right)^{-3}$	>>(1/2)^(-3)
123×10^{-4}	>>123*10^-4 or >>123e-4
$12^{\frac{3}{4}}$	>>12^(3/4)

Table 1.9: MATLAB's functions related to complex numbers

Function	Explanation	Example
complex	This command is used to produce a complex number, i.e., $complex(a,b) = a + b \times j$ where $j = \sqrt{-1}$.	>>complex(2,3) ans = 2.0000 + 3.0000i
angle	Returns the angle of complex number. It returns the angle in Radyans.	>>angle(2+4i) ans = 1.1071
conj	Returns the complex conjugate of a complex number.	>>conj(2+5i) ans = 2.0000 − 5.0000i
imag	Returns the imaginary part of a complex number.	>>imag(4+9i) ans = 9
real	Returns the real part of a complex number.	>>real(4+9i) ans = 4
isreal	When the input argument is a real number, i.e., imaginary part is zero, it returns 1. Otherwise it returns 0.	>>isreal(2+3i) ans = 0 >>isreal(4) ans = 1
abs	Calculates the absolut value. When the input argument is a complex number it returns the magnitude of the number, i.e., $abs(a + j \times b) = \sqrt{(a^2 + b^2)}$.	>>abs(3−4*i) ans = 5

Table 1.10: **MATLAB**'s rounding functions

Function	Explanation	Example
fix	Round toward zero.	>>fix(4.6) ans = 4
floor	Round toward negative infinity.	>>floor(8.4687) ans = 8
ceil	Round toward positive infinity.	>>ceil(4.4) ans = 5
round	Round to nearest decimal or integer.	>>round(4.55) ans = 5

Table 1.11: **MATLAB**'s remainder functions

Function	Explanation	Example
mod	Remainder after division (modulo operation).	>>mod(17,3) ans = 2
rem	Remainder after division.	>>rem(17,3) ans = 2

sign	Sign function (signum function).	>>sign(4)
		ans =
		        1
		>>sign(-4)
		ans =
		        -1

1.15 MATRIXES IN MATLAB®

MATLAB® considers everything as matrices. For instance, A, B, C, D are 1×1 matrices and E is a 1×14 matrix.

```
>>A=123;
```

```
>>B=-456.7;
```

```
>>C=12e-3;
```

```
>>D=12+34*i;
```

```
>>E='press any key...';
```

Table 1.12 shows how to define larger matrices.

Table 1.12: Entering a matrix in MATLAB®. (Note: in order to read the array in the 3rd row and 2nd column of matrix F, you write, F(3,2).)

Mathematics	MATLAB

$F = \begin{bmatrix} 1 & 2 & 3 \\ 4 & 5 & 6 \\ 7 & 8 & 9 \end{bmatrix}$	`>>F=[1 2 3;4 5 6;7 8 9]`
	 or
	 `>>F=[1,2,3;4,5,6;7,8,9]`

1.16 COMMANDS: RAND, ONES, ZEROS, EYE

- rand: produce a matrix with uniformly distributed random arrays. For instance, A=rand(1,3) makes a 1×3 matrix with random arrays. Each array is between 0 and 1.

- ones: reate array of all ones. For instance, ones(k), creates a $k \times k$ matrix which all the arrays are 1.

- zeros: create array of all zeros. For instance, zeros(k), creats a $k \times k$ matrix which all the arrays are 0.

- eye: makes identity matrix. For instance, eye(4) create a 4×4 identity matrix.

1.17   THE COLON OPERATOR

You can use the colon operator (:) to creat matrices. Here is the syntax:

$$f = \text{Initial value:Change steps:Final value}$$

Example 1.3

```
>>n=[1:2:10]

n =

      1 3 5 7 9

>>n(2:4)

ans =

      3 5 7
```

1.18   LOGSPACE AND LINSPACE COMMANDS

The `logspace` function generates logarithmically spaced vectors. For instance, `y=logspace(a,b)` generates a row vector y of 50 logarithmically spaced points between decades 10^a and 10^b. `y=logspace(a,b,n)` generates n points between decades 10^a and 10^b. `logspace` command is especially useful for creating frequency vectors.

The `linspace` function generates linearly spaced vectors. For instance, `y=linspace(a,b)` generates a row vector y of 50 linearly spaced points between decades a and b. `y=linspace(a,b,n)` generates n points between decades a and b.

1.19   OPERATION WITH MATRICES

In order to obtain the size of a matrix, use the `size` command. For instance, `size(A)` returns the number of rows and number of colums of a matrix named A.

`diag(A)` returns the arrays on the main diagonal of square matrix A.

`det(A)` returns the determinant of square matrix A.

`inv(A)` or A^-1 returns the inverse of square matrix A.

`eig(A)` returns the eigen values of square matrix A. `[v,d]=eig(A)` returns the eigen vectors as well.

`rank(A)` returns the rank of matrix A.

`transpose(A)` returns the transpose of matrix A.

`ctranspose(A)` or `A'` returns the transpose of matrix A. The complex conjugate transpose of a matrix interchanges the row and column index for each element, reflecting the elements across the main diagonal. The operation also negates the imaginary part of any complex numbers. For example, if `B = A'` and `A(1,2)` is 1+1i, then the element `B(2,1)` is 1-1i.

1.20   FINDING POLYNOMIAL ROOTS

Consider a polynomial function as follows:

$$f = x^4 + 2x^3 + 3x^2 + 4x + 5.$$

Enter the polynomial function coefficients as a vector f.

```
>>f=[1 2 3 4 5]
```

The roots can be find with the following command:

```
>>roots(f)
```

You can use the command `polyval` in order to obtain the value of a polynomial function at a specific point. For instance, in order to calculate the polynomial f at x=0 you write:

```
>>polyval(f,0)
```

1.21   PRODUCT OF TWO POLYNOMIALS

You can calculate the product of two polynomials with the aid of `conv` command. For instance, to find the product of $f = x^4 + 2x^3 + 3x^2 + 4x + 5$ and $g = x^3 + 5$, we write:

```
>>f=[1 2 3 4 5];
```

```
>>g=[1 0 0 5];
```

```
>>conv(f,g)
```

```
ans =

    1    2    3    9   15   15   20   25
```

So the product of afforementioned polynomials is, $x^7 + 2x^6 + 3x^5 + 9x^4 + 15x^3 + 15x^2 + 20x + 25$.

1.22 SOLUTION OF LINEAR SYSTEMS OF EQUATIONS

You can use the linear system of equations with the aid of MATLAB®. See Table 1.13.

Table 1.13: Solving a linear system of equations in MATLAB®. (Note: instead of B\A you can use inv(A)*B or A^-1*B.)

Mathematics	MATLAB
$$\begin{bmatrix} 1 & 2 & 3 \\ 4 & 5 & 6 \\ 7 & 8 & 9 \end{bmatrix} \times \begin{bmatrix} x \\ y \\ z \end{bmatrix} = \begin{bmatrix} 1 \\ 2 \\ 3 \end{bmatrix}$$	`>>A=[1 2 3;4 5 6;7 8 8];` `>>B=[1;2;3];` `>>B\A` `ans=` `    -0.3333` `     0.6667` `     0`

1.23 SOLUTION OF NONLINEAR SYSTEMS OF EQUATIONS

MATLAB® can solve the nonlinear system of equations as well. fzero function can be used for this purpose. The syntax is:

$$\text{fzero('Function Name',initial value)}$$

You must prepare a MATLAB® function that contains the desired equation. The structure of a MATLAB® function is shown bellow:

function *output_argument* = *FunctionName*(*InputArgument*)

output_argument = *DesiredEquation*

end

study the following example to see how to use the fzero command.

Example 1.4

Find the root of $f = x^4 + 3x^3 + 3x^2 + 14x + 5$ with the aid of `fzero` command.

Solution:

In order to prepare the MATLAB® function containing the non linear equations, enter the editor environment (Fig. 1.15).

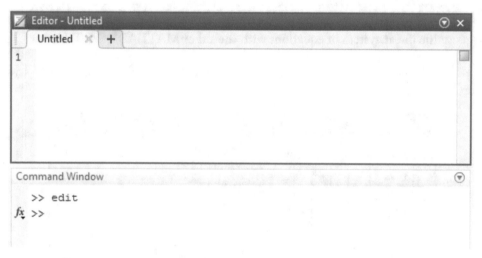

Figure 1.15: You can enter the editor by typing edit in the command line.

Write the following codes inside the editor (Fig. 1.16).

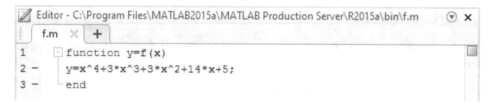

Figure 1.16: The function contains the equations.

```
function y=f(x)

y=x^4+3*x^3+3*x^2+14*x+5;

end
```

Save the file as f.m. When you save a function, you must use the same name as the function name. Since we call the MATLAB® function f, we must save it as f.m. Type the following command in command prompt of MATLAB®:

```
>>fzero(`f',0)

ans=

    -0.3776
```

NOTE: Set the "Current Folder" window of MATLAB® to where you saved f.m, otherwise MATLAB® cannot find the f.m.

1.24 SOLVING DIFFERENTIAL EQUATIONS

You can solve Ordinary Differential Equations (ODE) and even Partial Differential Equations (PDE) with the aid of MATLAB®.

See the following examples to learn different ways of solving ODEs.

Example 1.5
Solve $\frac{dy(t)}{dt} + 4y(t) = e^{-t}$, $y(0) = 1$ with MATLAB®.

Solution:
See Fig. 1.17.

```
Command Window                                              ⊙

    >> syms y(t)
    >> ode = diff(y)+4*y == exp(-t);
    >> cond = y(0) == 1;
    >> ySol(t) = dsolve(ode,cond)


    ySol(t) =

    exp(-t)/3 + (2*exp(-4*t))/3

fx >>
```

Figure 1.17: Solving the $\frac{dy(t)}{dt} + 4y(t) = e^{-t}$, $y(0) = 1$. (Note: syms defines symbolic variables.)

Example 1.6

Solve $2\frac{d^2y(t)}{dt^2} + \frac{dy(t)}{dt} + 11y(t) = e^{-t}$, $y(0) = 1$, $y'(0) = 0$ with MATLAB®.

Solution:

See Fig. 1.18.

```
Command Window

>> dsolve('2*D2y+D1y+11*y=exp(-t)','y(0)=1','D1y(0)=0')

ans =

(11*exp(-t/4)*cos((87^(1/2)*t)/4))/12 + (5*87^(1/2)*exp(-t/4)*sin((87^(1/2)*t)/4))/348 + (87^(1/2)*exp(-t/4)*exp(-(3*t)/4)*cos((87^(1

>> simplify(ans)

ans =

exp(-t)/12 + (11*exp(-t/4)*cos((87^(1/2)*t)/4))/12 + (5*87^(1/2)*exp(-t/4)*sin((87^(1/2)*t)/4))/348

fx >>
```

Figure 1.18: Solving the $2\frac{d^2y(t)}{dt^2} + \frac{dy(t)}{dt} + 11y(t) = e^{-t}$, $y(0) = 1$, $y'(0) = 0$.

This example used the "Dn" operator instead of "`diff`" command(Dn(.)= $\frac{d^n}{dt^n}(.)$).

Example 1.7

Solve $\frac{dy(t)}{dt} = ty$ with MATLAB®.

Solution:

See Fig. 1.19.

```
Command Window

>> syms y(t)
>> ode = diff(y,t) == t*y;
>> ySol(t) = dsolve(ode)

ySol(t) =

C8*exp(t^2/2)

fx >>
```

Figure 1.19: Solving the $\frac{dy(t)}{dt} = ty$.

Example 1.8

Solve $(\frac{dy(t)}{dt} + y(t))^2 = 1$, $y(0) = 0$ with MATLAB®.

Solution:

See Fig. 1.20.

```
Command Window
   >> syms y(t)
   >> ode = (diff(y,t)+y)^2 == 1;
   >> cond = y(0) == 0;
   >> ySol(t) = dsolve(ode,cond)

   ySol(t) =

    exp(-t) - 1
    1 - exp(-t)

fx >>
```

Figure 1.20: Solving the $(\frac{dy(t)}{dt} + y(t))^2 = 1$, $y(0) = 0$.

Example 1.9

Solve $\frac{d^2y(x)}{dx^2} = \cos(2x) - y$, $y(0) = 1$, $y'(0) = 0$ with MATLAB®.

Solution:

See Fig. 1.21.

```
Command Window                                                    ⊙

   >> syms y(x)
   >> Dy = diff(y);
   >> ode = diff(y,x,2) == cos(2*x)-y;
   >> cond1 = y(0) == 1;
   >> cond2 = Dy(0) == 0;
   >> conds = [cond1 cond2];
   >> ySol(x) = dsolve(ode,conds);
   >> ySol = simplify(ySol)


   ySol(x) =


   1 - (8*sin(x/2)^4)/3

fx >>
```

Figure 1.21: Solving the $\frac{d^2 y(x)}{dx^2} = \cos(2x) - y$, $y(0) = 1$, $y'(0) = 0$.

1.25   DIFFERENTION AND INTEGRATION

Examples for differention and integration in MATLAB® are shown in Table 1.14.

Table 1.14: Some examples for differention and integration in MATLAB®

Mathematics	MATLAB
$f(x) = \dfrac{(x^3 - x)}{(x^2 + 8)}$ $\dfrac{df}{dx} = ?$	`>>syms x` `>>diff('(x^3-x)/(x^2+8)')`
$f(x,y) = x^3 + 3y^2 + 4xy$ $\dfrac{df}{dy} = ?$	`>>syms x y` `>>f=x^3+3*y^2+4*x*y` `>>diff(f,y)`
$\displaystyle\int \dfrac{x^3 - x}{x^2 + 8}\,dx$	`>>syms x` `>>int((x^3-x)/(x^2+8))`
$\displaystyle\int_0^1 \dfrac{x^3}{x^2 + 8}\,dx$	`>>syms x` `>>int((x^3)/(x^2+8),x,0,1)`

1.26 DRAWING DIAGRAMS WITH MATLAB®

You can use the plot(x,y) command to draw diagrams. See the following examples.

Example 1.10

```
>>x=[0:0.01:2*pi];

>>y=sin(x);

>>plot(x,y)
```

See Fig. 1.22.

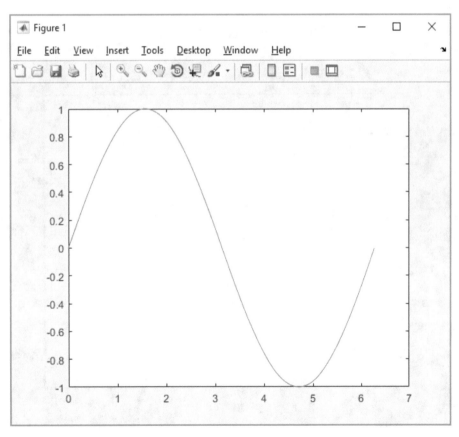

Figure 1.22: Output of Example 1.10.

Example 1.11

```
>>x=[0:0.01:2*pi];

>>y1=sin(x);

>>y2=cos(x);

>>plot(x,y1,x,y2)
```

See Fig. 1.23.

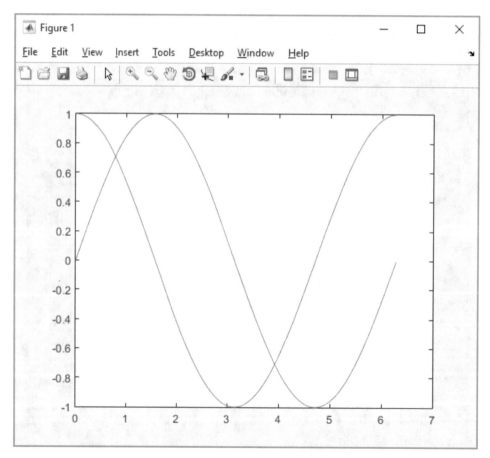

Figure 1.23: Output of Example 1.11.

You can select the desired line style, color or marker type for the drawn diagram. To do this, you use the `plot(x,y,'?')` command. ? shows the desired line style/color/marker type. Line style, color or marker type is set according to Tables 1.15, 1.16, and 1.17.

Table 1.15: Line style table

Command	Line style
-	Solid line
--	Dashed line
:	Dotted line
-.	Dotted and dashed line

Table 1.16: Colors table

Command	Color
r	red
g	green
b	blue
c	cyan
m	magenta
y	yellow
k	black
w	white

Table 1.17: Available markers

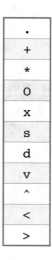

Table 1.18 shows the commands in relation to the `plot` command.

Table 1.18: Some of the commands related to `plot` command

Command	Explanation
grid	Display or hide axes grid lines
title('')	Add title.
xlabel('')	Label X axis.
ylabel('')	Label Y axis.
text(x,y,'')	Add text descriptions to data points

See the following examples.

Example 1.12

```
>>x=[0:0.01:2*pi];

>>y=sin(x+pi/4);

>>plot(x,y,'b--')

>>xlabel(`Time(s)')
```

```
>>ylabel(`Voltage Values(V)')

>>grid on
```

See Fig. 1.24.

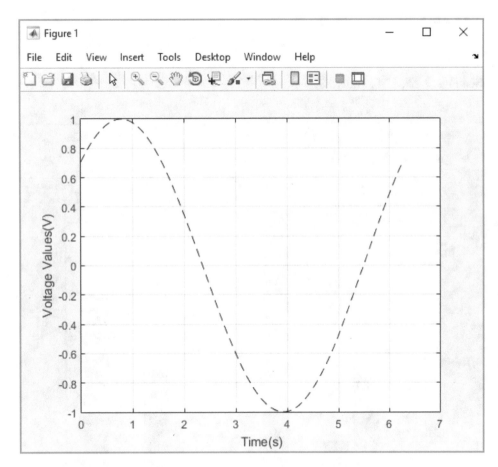

Figure 1.24: Output of Example 1.12.

You can draw the bargraph, stairstep graph, and pie chart with the aid of bar, stair, and pie commands, respectively.

Example 1.13

```
>>a=[0 1 3 7 9];
```

```
>>b=[2 1 4 2 3];

>>figure(1)

>>bar(a,b)

>>stairs(a,b)

>>pie(a)
```

See Figs. 1.25, 1.26, and 1.27.

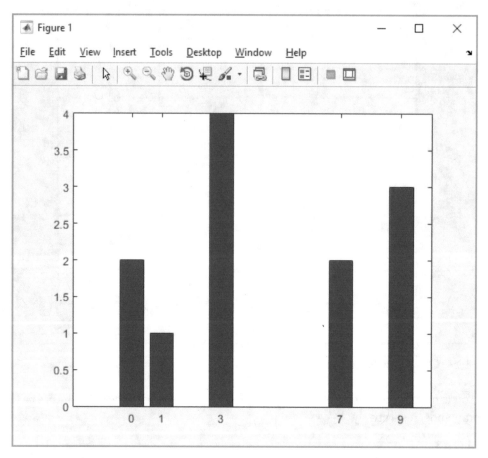

Figure 1.25: Output of Example 1.13.

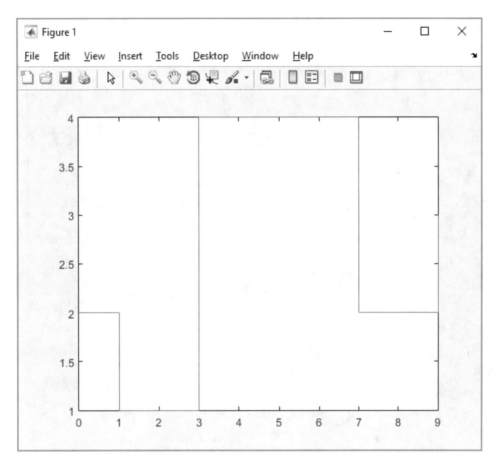

Figure 1.26: **Output of Example** 1.13.

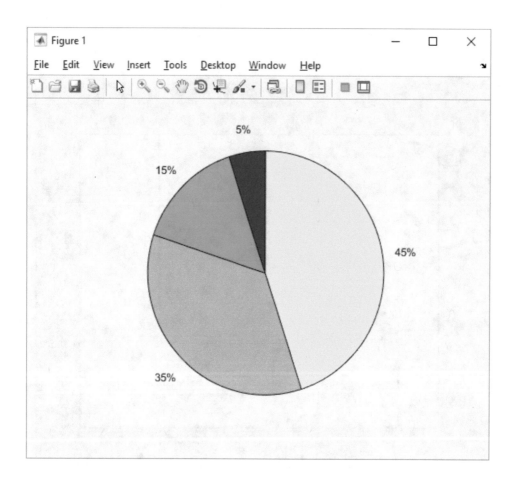

Figure 1.27: **Output of Example** 1.13.

1.27 DRAWING THE GRAPH OF COLLECTED DATA

MATLAB® can be used to draw the graph of data collected in lab. Both linear and logarithmic graphs can be draw with the aid of MATLAB®. The following example shows how to do this.

Example 1.14

Table 1.19 shows the current-voltage for a resistor. We want to draw the plot of this data.

Table 1.19: Measured voltage and current for resistor R_1

V (Volt)	I (Amper)
0.499	0.10
0.985	0.20
1.508	0.31
1.969	0.41
2.528	0.53
2.935	0.61
3.481	0.73
3.971	0.83
4.486	0.94
4.960	1.04
5.502	1.15
6.007	1.26
6.60	1.38

Enter the obtained data to MATLAB®. See Fig. 1.28.

```
Command Window                                                                    ⊙
  >> V=[.499 .985 1.508 1.969 2.528 2.935 3.481 3.971 4.486 4.960 5.502 6.007 6.600];
  >> I=[.1 .2 .31 .41 .53 .61 .73 .83 .94 1.04 1.15 1.26 1.38];
fx >>
```

Figure 1.28: Entering the measured data.

Use the plot command to draw the graph (Fig. 1.29). The result is shown in Fig. 1.30.

If you need smaller grid axes, use the command grid minor. Results are shown in Figs. 1.31 and 1.32.

Command Window

```
>> V=[.499 .985 1.508 1.969 2.528 2.935 3.481 3.971 4.486 4.960 5.502 6.007 6.600];
>> I=[.1 .2 .31 .41 .53 .61 .73 .83 .94 1.04 1.15 1.26 1.38];
>> plot(V,I),grid on
>>
```

Figure 1.29: **Drawing the data.**

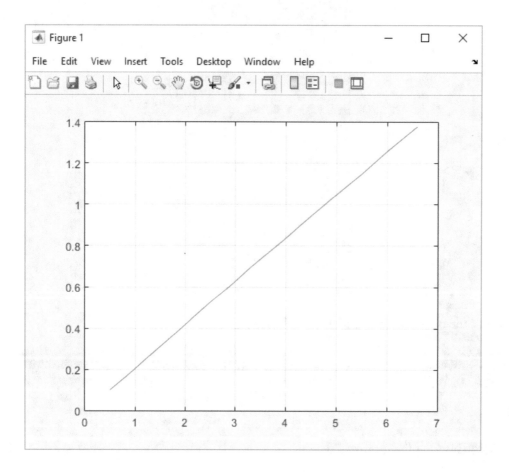

Figure 1.30: **Graph of measured data.**

```
Command Window
>> V=[.499 .985 1.508 1.969 2.528 2.935 3.481 3.971 4.486 4.960 5.502 6.007 6.600];
>> I=[.1 .2 .31 .41 .53 .61 .73 .83 .94 1.04 1.15 1.26 1.38];
>> plot(V,I),grid on
>> plot(V,I),grid minor
fx >>
```

Figure 1.31: Using smaller grid axes.

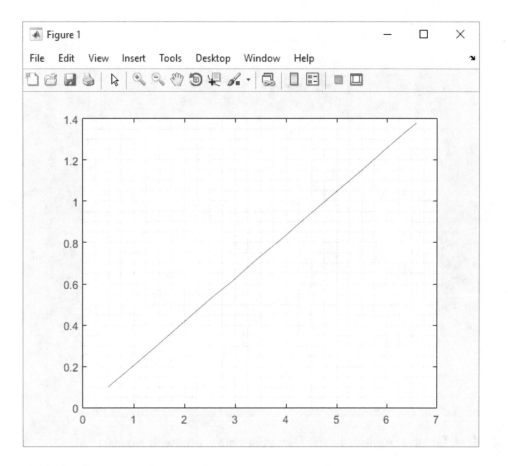

Figure 1.32: Smaller axes grids are used.

You can use the command shown in Fig. 1.33 in order to show the data points with a marker. The result is shown in Fig. 1.34.

```
Command Window
>> V=[.499 .985 1.508 1.969 2.528 2.935 3.481 3.971 4.486 4.960 5.502 6.007 6.600];
>> I=[.1 .2 .31 .41 .53 .61 .73 .83 .94 1.04 1.15 1.26 1.38];
>> plot(V,I),grid on
>> plot(V,I),grid minor
>> plot(V,I,'r*',V,I),grid minor
fx >>
```

Figure 1.33: Showing the data points with red stars.

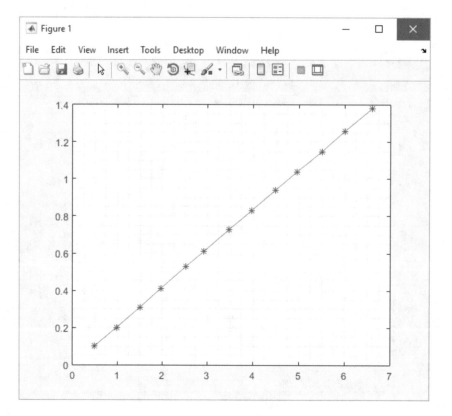

Figure 1.34: Graph of measured data. Data points are shown with red stars.

Use "`xlabel`" and "`ylabel`" to add descriptive texts to the axes (Figs. 1.35 and 1.36).

```
Command Window
   >> V=[.499 .985 1.508 1.969 2.528 2.935 3.481 3.971 4.486 4.960 5.502 6.007 6.600];
   >> I=[.1 .2 .31 .41 .53 .61 .73 .83 .94 1.04 1.15 1.26 1.38];
   >> plot(V,I),grid on
   >> plot(V,I),grid minor
   >> plot(V,I,'r*',V,I),grid minor
   >> xlabel('Resistor Voltage(V)')
   >> ylabel('Resistor Current(A)')
fx >>
```

Figure 1.35: Adding labels to axes.

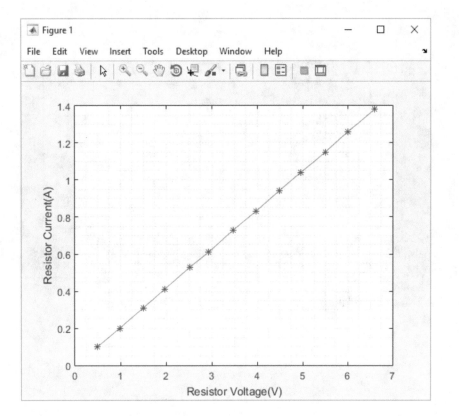

Figure 1.36: Labels are added to axes.

You can add text to axes with the aid of **Insert** menu (see Fig. 1.37) as well. After Insert menu is opened, click on "X Label" or "Y Label" to add text to x axis and y axis, respectively.

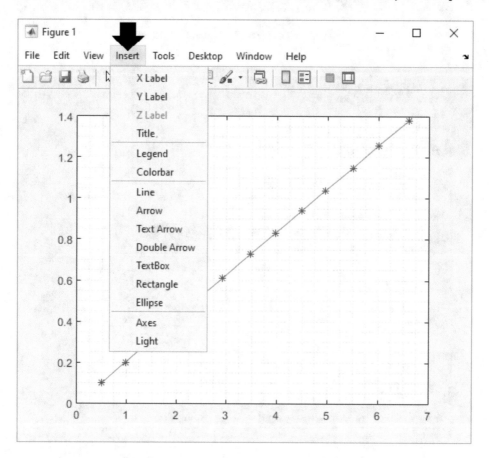

Figure 1.37: You can add labels to the axes with the aid of insert menu.

You can save the drawn graph as a graphic file. To do this use **File>Save As...** See Fig. 1.38.

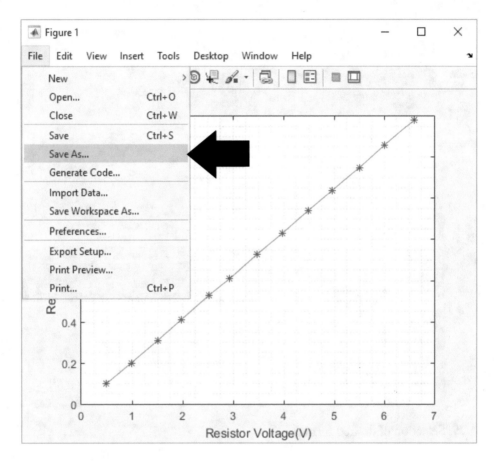

Figure 1.38: You can export the drawn graph as a graphic file.

Select the desired graphic file format and save the file. See Fig. 1.39.

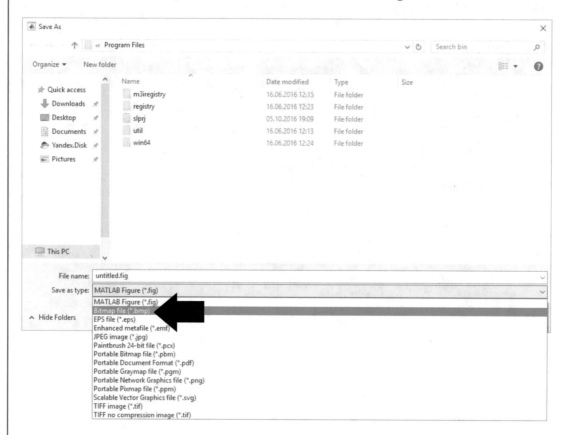

Figure 1.39: Selection of desired format.

1.28 DRAWING TWO OR MORE DATA SIMULTANOUSLY

Drawing a graph of two or more data simultanously is so common. Drawing such graphs in MATLAB® is shown with an example.

Consider we have measured the data shown in Table 1.20 for another resistor.

We want to show the I–V relationship of the first resistor (Table 1.19) and the second resistor (Table 1.20) on the same graph. Codes shown in Fig. 1.40 show how to do this. The result is shown in Fig. 1.41.

Use **Insert>Legend** to show which plot belong to which resistor (Fig. 1.42).

After you clicked the "Legend," MATLAB® adds the default legend to the plot (Fig. 1.43).

Double-click the data1, data2, data3, and data4 to change them into a more convinient text. You can drag the legend box to desired place, as well (Fig. 1.44).

Table 1.20: Measured voltage and current for resistor R_2

V (Volt)	I (Amper)
0.579	0.10
0.978	0.17
1.598	0.28
1.976	0.34
2.496	0.43
2.953	0.51
3.458	0.60
4.068	0.71
4.450	0.78
4.917	0.86
5.35	0.93
5.75	1.01
6.37	1.11
6.60	1.15

```
Command Window
  >> I1=[.1 .2 .31 .41 .53 .61 .73 .83 .94 1.04 1.15 1.26 1.38];
  >> V1=[.499 .985 1.508 1.969 2.528 2.935 3.481 3.971 4.486 4.960 5.502 6.007 6.600];
  >> I2=[.1 .17 .28 .34 .43 .51 .60 .71 .78 .86 .93 1.01 1.11 1.15];
  >> V2=[.579 .978 1.598 1.976 2.496 2.953 3.458 4.068 4.450 4.917 5.35 5.75 6.37 6.6];
  >> plot(V1,I1,'*r',V1,I1)
  >> hold on
  >> plot(V2,I2,'+r',V2,I2)
  >> grid minor
fx >> |
```

Figure 1.40: You draw two or more plots on the same axes simultanously with the aid of hold on command.

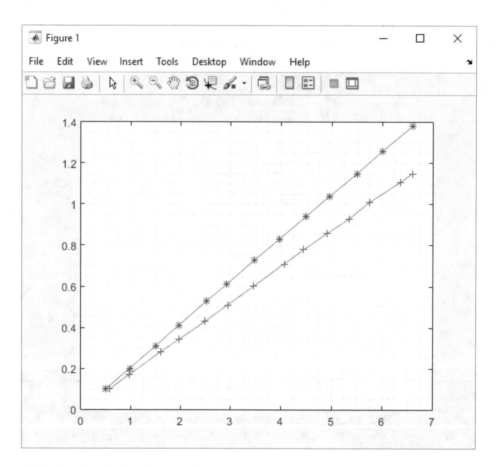

Figure 1.41: **Result of code shown in Fig.** 1.40.

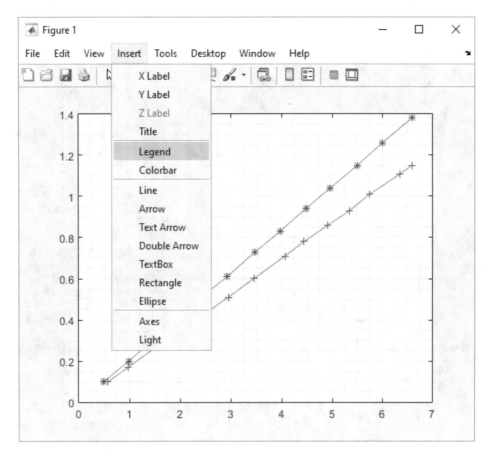

Figure 1.42: Adding legend to the drawn graph.

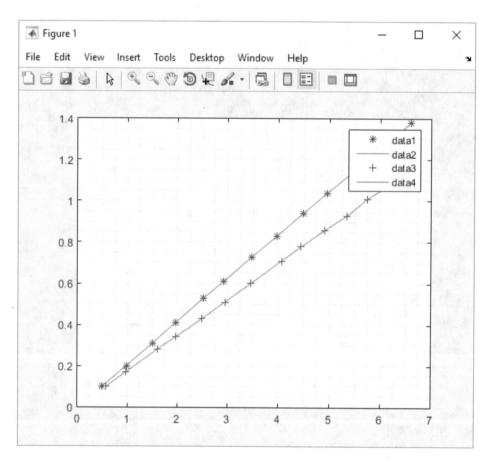

Figure 1.43: Default legend is added to the graph.

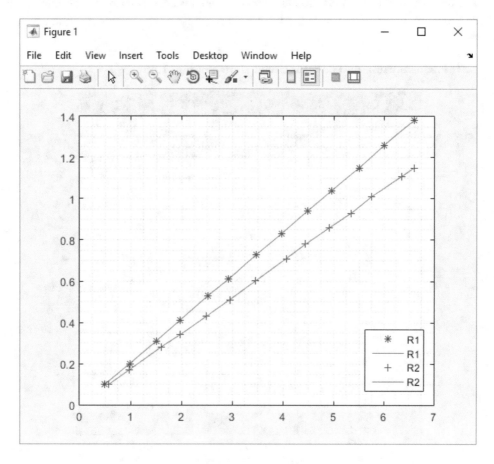

Figure 1.44: Editing and replacing the default legend.

1.29   DRAWING THE FREQUENCY RESPONSE

Frequency response plots are drawn on the logarithmic scale. You can draw the logarithmic plots with the aid of `semilogx`, `semilogy`, and `loglog` commands.

Drawing the frequency response plot is usually done with the aid of `semilogx` command. Since only the x axis of frequency response is in logarithmic scale.

The procedure of drawing frequency response plots is shown with an example. Consider a data as shown in Table 1.21. This is the frequency response of the simple RC circuit shown in Fig. 1.45.

Table 1.21: Frequency response data for circuit shown in Fig. 1.45

Frequency (Hz)	Magnitude $\left(\left\|\dfrac{V_o\,(j\omega)}{V_{in}\,(j\omega)}\right\|\right)$	Phase $\left(\sphericalangle\dfrac{V_o\,(j\omega)}{V_{in}\,(j\omega)}\right)$ (Degrees)
1	1.000	-0.36
10	0.998	-3.60
20	0.992	-7.16
50	0.954	-17.44
100	0.847	-32.13
150	0.728	-43.30
200	0.623	-51.48
250	0.537	-57.51
300	0.469	-62.05
350	0.414	-65.54
400	0.370	-68.30
450	0.333	-70.51
500	0.303	-72.34
550	0.278	-73.85
600	0.256	-75.14

We enter the data to MATLAB® (Fig. 1.46).

We want to draw the response in dB. To do this we must calculate the $20\log_{10}\left(\left|\frac{V_o(j\omega)}{V_{in}(j\omega)}\right|\right)$. Commands shown in Fig. 1.47 draw the frequency response of data given in Table 1.21. The result is shown in Fig. 1.48.

We need to draw the phase diagram as well. We add the command shown in Fig. 1.49 to draw the phase diagram. The result is shown in Fig. 1.50.

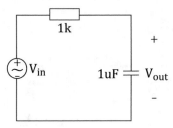

Figure 1.45: Simple lowpass RC circuit.

```
Command Window                                                                              ⊙
   >> f=[1 10 20 50 100 150 200 250 300 350 400 450 500 550 600];
   >> magntd=[1 .998 .992 .954 .847 .728 .623 .537 .469 .414 .370 .333 .303 .278 .256];
   >> phz=[-.36 -3.6 -7.16 -17.44 -32.13 -43.30 -51.48 -57.51 -62.05 -65.54 -68.30 -70.51 -72.34 -73.85 -75.14];
 fx >> |
```

Figure 1.46: Entering the magnitude and phase to MATLAB®.

```
Command Window                                                                              ⊙
   >> f=[1 10 20 50 100 150 200 250 300 350 400 450 500 550 600];
   >> magntd =[1 .998 .992 .954 .847 .728 .623 .537 .469 .414 .370 .
   >> phz =[-.36 -3.6 -7.16 -17.44 -32.13 -43.30 -51.48 -57.51 -62.0!
   >> subplot(211),semilogx(f,20*log10(magntd)),grid minor
```

Figure 1.47: Drawing the magnitude graph.

You can add a descriptive title to the graph. To do this, click the **Insert>Title** and type the desired title (Fig. 1.51).

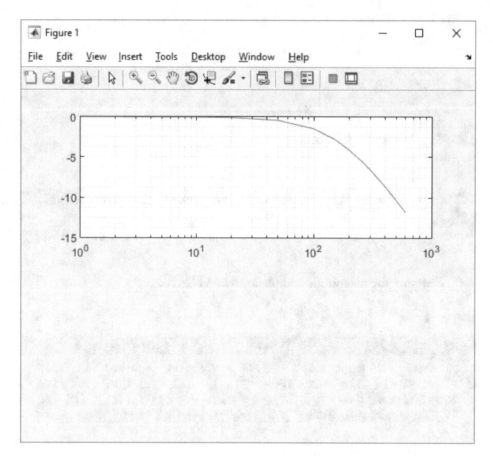

Figure 1.48: The magnitude graph.

```
Command Window                                                        ⊙
  >> f=[1 10 20 50 100 150 200 250 300 350 400 450 500 550 600];
  >> magntd=[1 .998 .992 .954 .847 .728 .623 .537 .469 .414 .370 .
  >> phz=[-.36 -3.6 -7.16 -17.44 -32.13 -43.30 -51.48 -57.51 -62.0
  >> subplot(211),semilogx(f,20*log10(magntd)),grid minor
  >> subplot(212),semilogx(f,phz),grid minor
fx >> |
```

Figure 1.49: Drawing the phase graph.

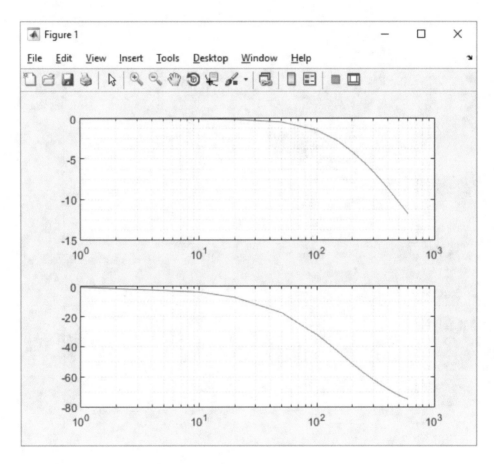

Figure 1.50: The magnitude and phase graph.

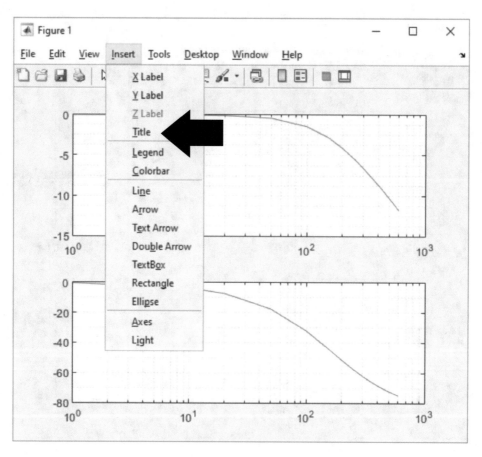

Figure 1.51: Addition of a title to the drawn plot.

CHAPTER 2

Commonly Used Commands in Analysis of Control Systems

2.1 INTRODUCTION

Chapter 1, was a general introduction to MATLAB. This chapter, introduces some of the important commands used in analysis of control systems.

2.2 DEFINING THE TRANSFER FUNCTION

The procedure is shown with two examples.

Example 2.1
We want to define the following transfer function in MATLAB.

$$H(s) = \frac{100}{s^2 + 6s + 100}$$

```
>>H=tf([100],[1 6 100]);
```

As another way, you can write:

```
>> s=tf('s');
```

```
>> H=100/(s^2+6*s+100)
```

Example 2.2
To define following transfer function,

$$I(s) = \frac{s^2 - 6s}{s^3 + 6s + 100}$$

```
>>I=tf([100 -6 0],[1 0 6 100]);
```

Example 2.3

To define the following transfer function,

$$J(s) = \frac{s+1}{(s+2)\,(s+3)\,(s+4)}.$$

One may do some simple manipulation and write,

$$J(s) = \frac{s+1}{(s+2)\,(s+3)\,(s+4)} = \frac{s+1}{s^3 + 9s^2 + 26s + 24},$$

and then use the following command:

```
>>J=tf([1 1],[1 9 26 24]);
```

However there is a simpler solution using zpk command:

```
>>J=zpk([-1],[-2 -3 -4],1);
```

The zpk command takes three inputs: the first input is a vector contain transfer function zeros, the second is the vector contains transfer function poles, and the third one is a gain. Since J, has only one zero places at -1, the zero vector contains only one array, -1. The J has poles at -2, -3, and -4. So, second input of zpk command is $[-2\ -3\ -4]$. The gain zero since $J(s) = \frac{s+1}{(s+2)(s+3)(s+4)} = 1 \times \frac{s+1}{(s+2)(s+3)(s+4)}$. To model $K(s) = -5\frac{s-1}{(s+2)(s-3)(s+4)}$ one must use the command >>K=zpk([+1],[-2 +3 -4],-5).

Figure 2.1 shows how to transfer a zpk model to tf model.

You can define state space models as well. Use ss command to define state space models. See Fig. 2.2 to see an example.

You can obtain the transfer function associated with the state space model with the aid of tf command (Fig. 2.3).

You can obtain the location of polse and zeros with the aid of pole and zero commands, respectively (Figs. 2.4 and 2.5).

You can obtain a graphical plot of location of poles and zeros with the aid of pzmap command. pzmap shows the location of poles with a × mark and location of zeors with an open circle. For instance, you can draw the pole-zero map of L with the aid of the following command. The result is shown in Fig. 2.6.

```
>>pzmap(L)
```

A system is stable if all the poles lies in the open Left Half Plane (LHP), i.e., the left half plane without the imaginary access. You can use the command isstable to learn whether or not the system is stable.

```
Command Window                                          ⊙
   >> J=zpk([-1],[-2 -3 -4],1)

   J =

              (s+1)
     ------------------
      (s+2)  (s+3)  (s+4)

   Continuous-time zero/pole/gain model.

   >> tf(J)

   ans =

                  s + 1
     -----------------------
      s^3 + 9 s^2 + 26 s + 24

   Continuous-time transfer function.

fx >>
```

Figure 2.1: Converting a zpk model to tf model.

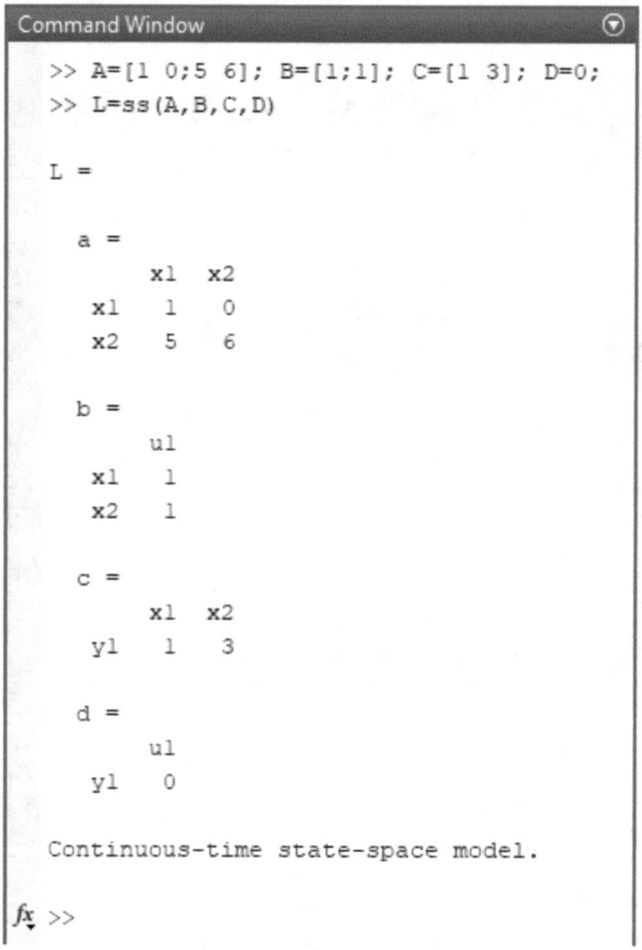

```
Command Window                                    ⊙
  >> A=[1 0;5 6]; B=[1;1]; C=[1 3]; D=0;
  >> L=ss(A,B,C,D)

  L =

    a =
          x1   x2
      x1   1   0
      x2   5   6

    b =
          u1
      x1   1
      x2   1

    c =
          x1   x2
      y1   1   3

    d =
          u1
      y1   0

  Continuous-time state-space model.

fx >>
```

Figure 2.2: Defining the state space model.

```
Command Window                                    ⊙
  >> A=[1 0;5 6]; B=[1;1]; C=[1 3]; D=0;  ∧
  >> L=ss(A,B,C,D)

  L =

    a =
          x1   x2
      x1   1    0
      x2   5    6

    b =
          u1
      x1   1
      x2   1

    c =
          x1   x2
      y1   1    3

    d =
          u1
      y1   0

  Continuous-time state-space model.

  >> tf(L)

  ans =

        4 s + 6
      -------------
      s^2 - 7 s + 6

  Continuous-time transfer function.

fx >> |                                           ∨
```

Figure 2.3: Converting a state space model to transfer function model.

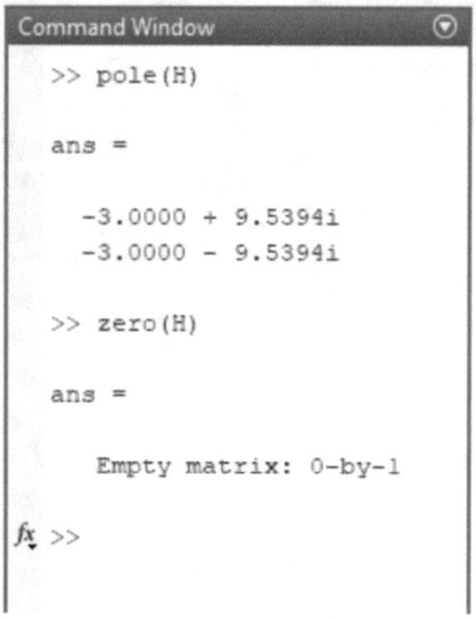

Command Window
>> pole(H)
ans =
-3.0000 + 9.5394i
-3.0000 - 9.5394i
>> zero(H)
ans =
Empty matrix: 0-by-1
fx >>

Figure 2.4: Obtaining the poles and zeros. `H`, is a transfer function model.

Command Window
>> pole(L)
ans =
6
1
>> zero(L)
ans =
-1.5000
fx >>

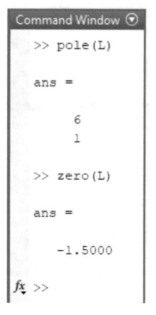

Figure 2.5: Obtaining the poles and zeros. `L`, is a state space model.

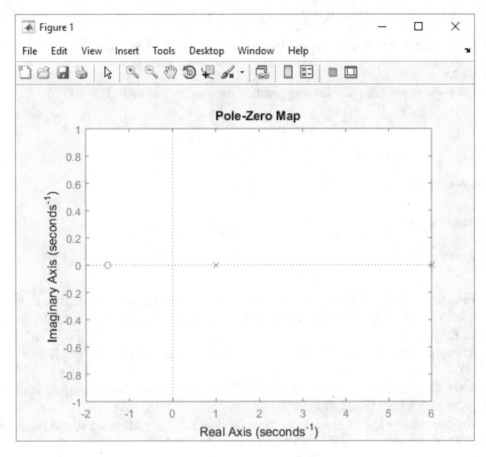

Figure 2.6: Pole-zero map of L.

```
>>H=tf([100],[1 6 100]);

>>isstable(H)

ans=

    1
```

2.3   IMPULSE RESPONSE OF A SYSTEM

Impluse response of a system can be drawn with the aid of `impulse` command.

Example 2.4

Draw the impulse response of following system. The result is shown in Fig. 2.7.

$$H(s) = \frac{100}{s^2 + 6s + 100}$$

```
>>H=tf([100],[1 6 100]);

>>impluse(H)

>>grid on
```

You can ask MATLAB to show peak response and settling time. To do this, right-click on the plot. Click on the Characteristics and select your desired option. For instance, we select the Peak Response (Fig. 2.8).

After you clicked the Peak Response, the MATLAB shows the Peak of graph with a point (Fig. 2.9).

After you clicked on the shown point, MATLAB adds a label and shows the coordimnate of the point with a label (Fig. 2.10).

You can use the Data Cursor icon shown in Fig. 2.11 to add cursor to the graph. First click the Data Cursor icon and then click a point on the graph.

Using the cursor, you can read the coordinate of desired point easily (Fig. 2.12).

You can slide the cursor to desired point. You can slide the cursor more easily if you set the Selection Style to Mouse Position (Fig. 2.13). To see the Selection Style, right-click on the graph.

You can zoom in/out the drawn graph with the aid of a magnifier icon (Fig. 2.14).

You can export the drawn graph as a graphic file. To do this, click theSave as… (Fig. 2.15).

Type your desired name and Select the desired graphic file (i.e., bmp, jpg, png) and export the graphic (Fig. 2.16).

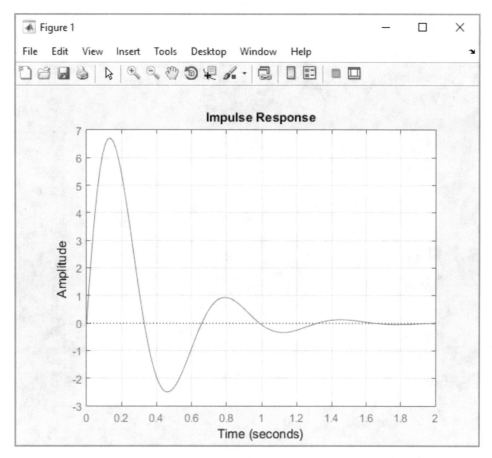

Figure 2.7: Impulse response of H.

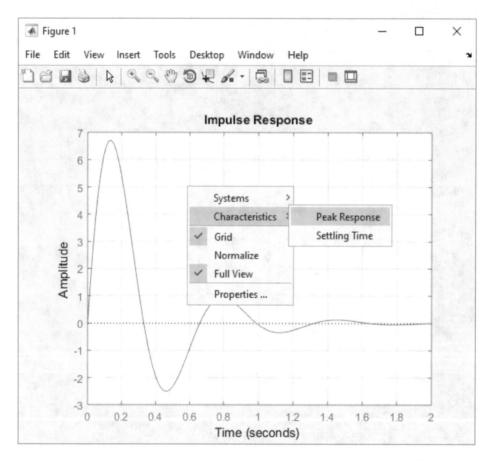

Figure 2.8: Obtaining the impulse response characteristics.

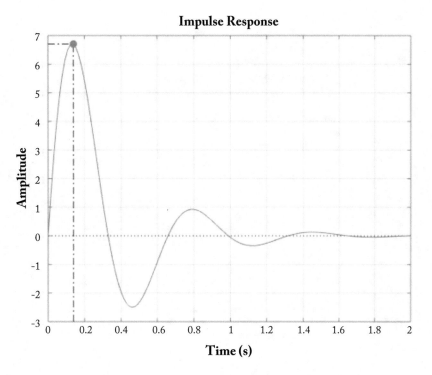

Figure 2.9: Peak of the graph is shown with a filled blue circle.

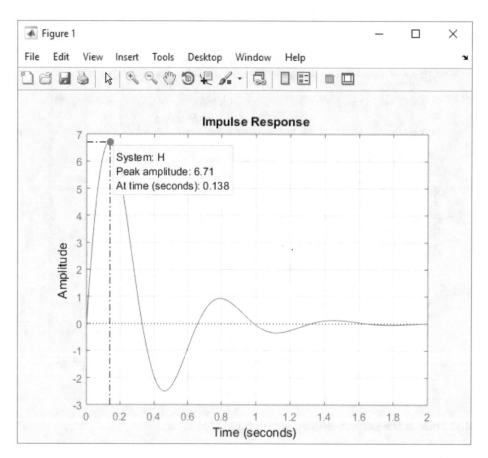

Figure 2.10: Coordinate of the peak point.

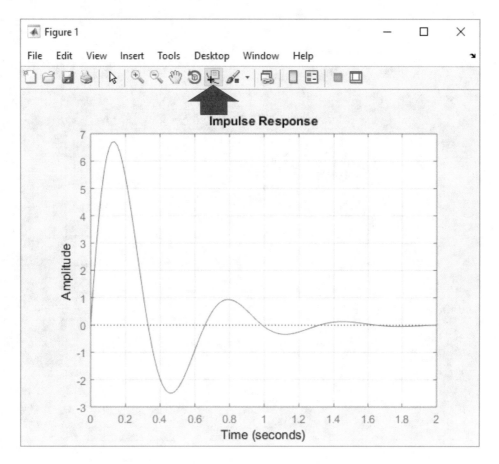

Figure 2.11: Data cursor icon.

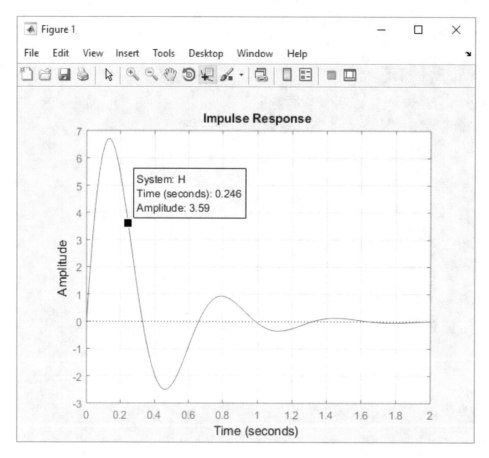

Figure 2.12: You can read the coordinate of desired point with the aid of added cursor. To remove the shown box, left-click on it and press the delete key on the keyboard.

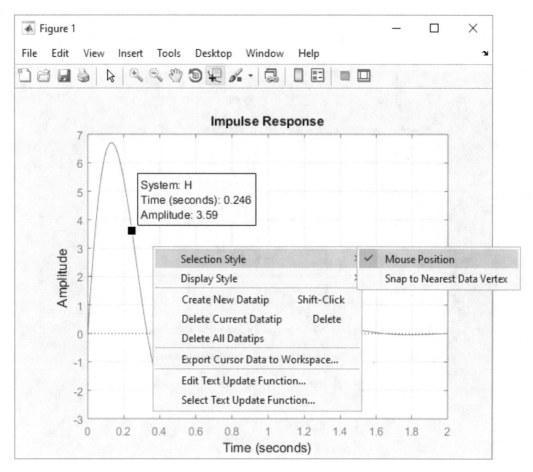

Figure 2.13: You can move the cursor more smoothly if you select the mouse position.

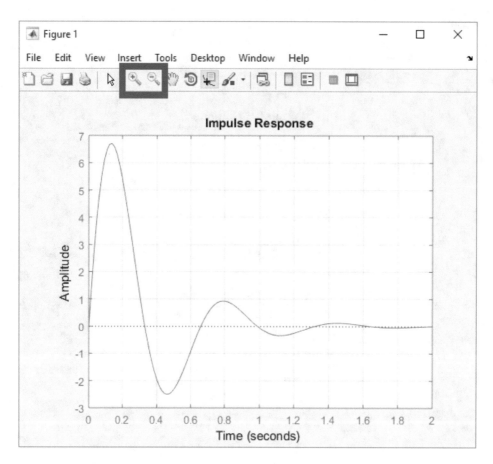

Figure 2.14: You can zoom in/out with the aid of a magnifier.

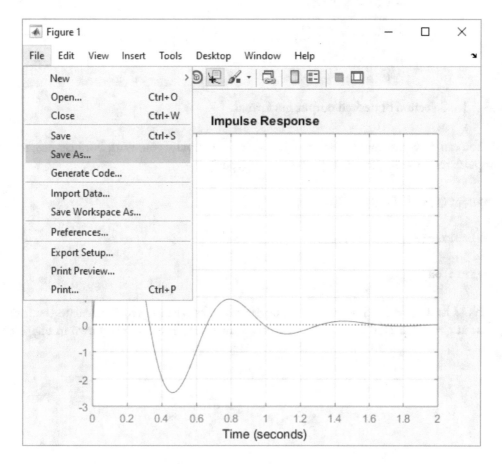

Figure 2.15: You can save the drawn plot with the aid of save as ....

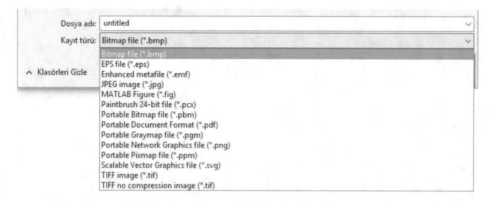

Figure 2.16: Selection of desired output file format.

You can draw the impulse response in a specific interval if you want. For instance, to draw the impulse response in the [.2,1.8] interval you type:

```
>>H=tf([100],[1 6 100]);
```

```
>>impluse(H,[.2:.01:1.8])
```

```
>>grid on
```

NOTE: .01, is the time step. The above code asks MATLAB to calculate the impulse response at $t = 0.2, 0.21, 0.22, 0.23, \ldots, 1.8$ and draw them. The result is shown in Fig. 2.17.

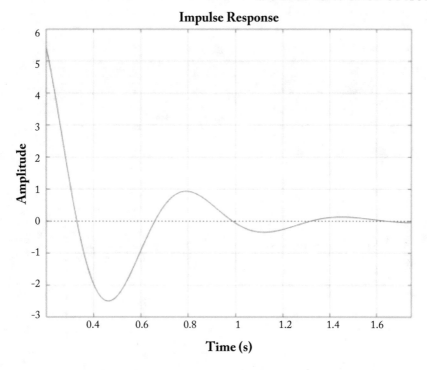

Figure 2.17: Drawing the impulse response for [0.2, 1.8].

2.4   STEP RESPONSE OF A SYSTEM

Step response of a system can be drawn with the aid of `step` command.

Example 2.5
Draw the step response of following system. The result is shown in Fig. 2.18.

$$H(s) = \frac{100}{s^2 + 6s + 100}$$

```
>>H=tf([100],[1 6 100]);

>>step(H)

>>grid on
```

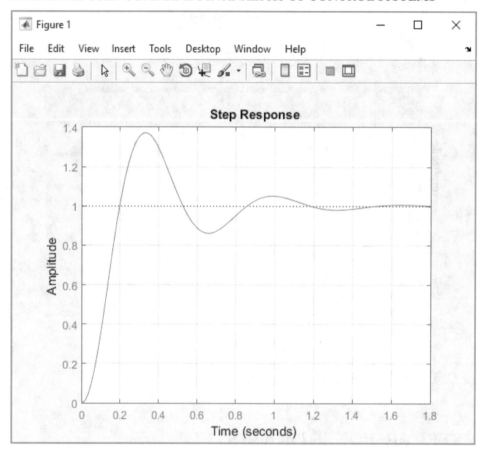

Figure 2.18: Step response of H.

You can draw the step response in a specific interval if you want. For instance, to draw the step response in the [.2,1.8] interval you type:

```
>>H=tf([100],[1 6 100]);

>>step(H,[.2:.01:1.8])

>>grid on
```

NOTE: .01, is the time step. The above code asks MATLAB to calculate the step response at $t = 0.2, 0.21, 0.22, 0.23, \ldots, 1.8$ and draw them. The result is shown in Fig. 2.19.

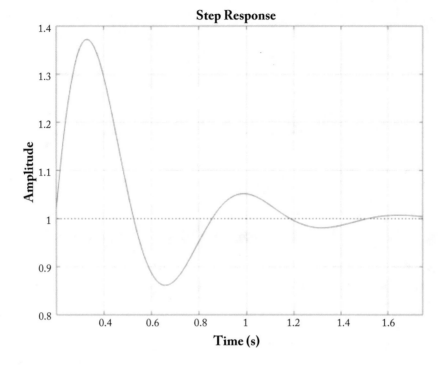

Figure 2.19: Drawing the impulse response for [0.2,1.8].

You can ask MATLAB to measure the important characteristics of the response and shows them. To do this, right-click on the step response plot and click characteristics. Select desired option from the shown menu. See Fig. 2.20.

For instance, we select the Rise Time from the shown menu. The result is shown in Fig. 2.21. Rise time is the time it takes for the response to rise from 10–90% of the steady-state response.

You can use the command `stepinfo` to see the step response characteristics (Fig. 2.22).

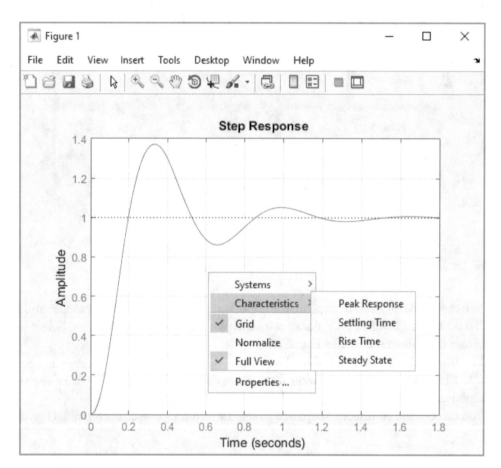

Figure 2.20: Obtaining the characteristics of the drawn graph.

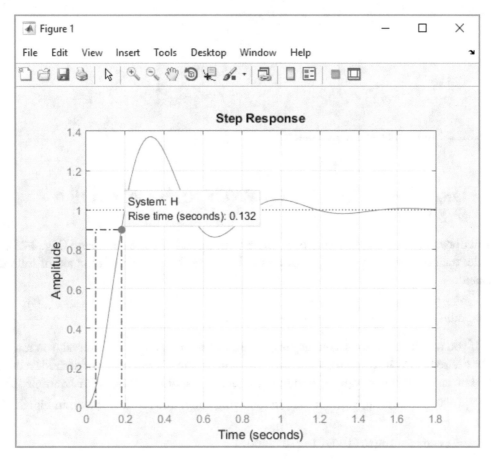

Figure 2.21: The rise time of the system is 0.132.

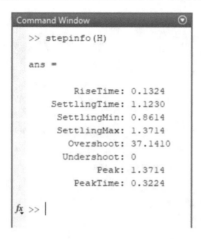

Figure 2.22: Step response characteristics.

2.5   DRAWING THE FREQUENCY RESPONSE OF A SYSTEM

You can draw the frequency response of a system with the aid of bode command (Fig. 2.23). For example, the bode diagram of system of Example 2.1 can be drawn with the aid of following command:

```
>>bode(H),grid on
```

If you want to draw the frequency response of a system in a specific interval, you must define the desired interval. You can use the bode(sys,w) where w is the vector of desired frequency range. For instance, if you want to draw the frequency response of system in Example 2.1 for $\left[100\frac{Rad}{s}, 1000\frac{Rad}{s}\right]$ interval, you can use the following code. The result is shown in Fig. 2.24.

```
>>w=logspace(log10(100),log10(1000))
```

```
>>bode(H,w)
```

```
>>grid on
```

You can measure the important characteristics of the drawn graph by right-clicking on it and select the Characteristics (Fig. 2.25).

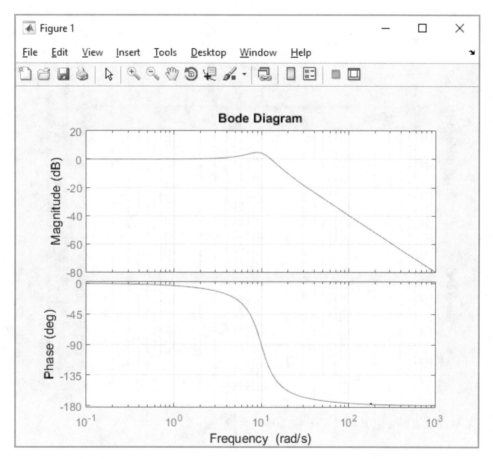

Figure 2.23: Bode plot of H.

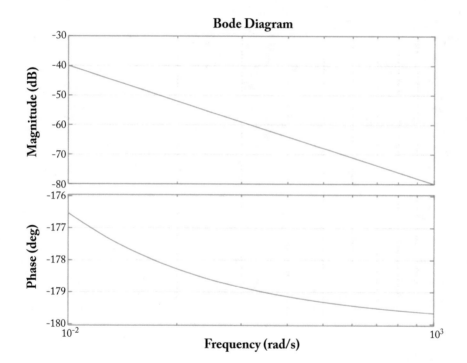

Figure 2.24: Bode plot of H for $\left[100\frac{\text{Rad}}{s}, 1000\frac{\text{Rad}}{s}\right]$ interval.

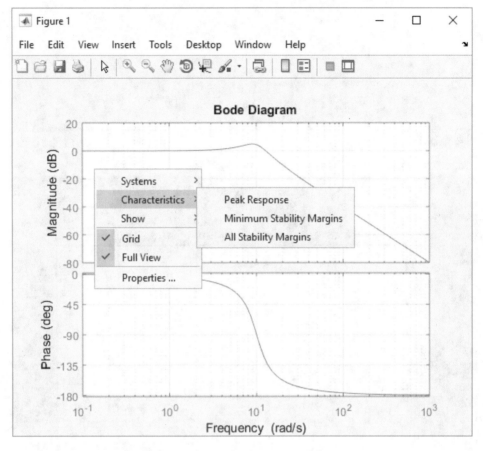

Figure 2.25: Obtaining the characteristics of the plot.

2.6   DRAWING THE NYQUIST DIAGRAM OF A SYSTEM

You can draw the Nyquist diagram of a system with the aid of `nyquist` command. For example, the Nyquist diagram of system of Example 2.1 can be drawn with the aid of following command. The result is shown in Fig. 2.26.

```
>>nyquist(H)
```

You can see the characteristics of the drawn plot by right-clicking the plot and select the characteristics (Fig. 2.27).

You can add the grids to the drawn graph by right-clicking on the graph and click the Grid in the appeared list (Fig. 2.28). The result is shown in Fig. 2.29. Another way of adding grid to your graph is typing the `>>grid on` in the MATLAB command prompt.

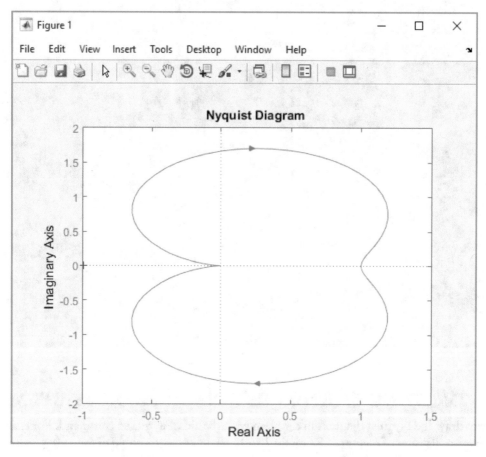

Figure 2.26: The Nyquist plot of the H.

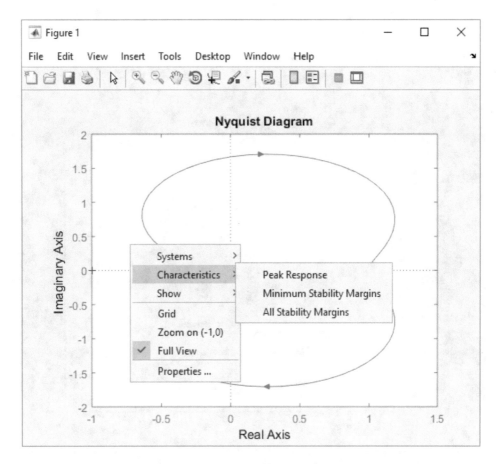

Figure 2.27: Obtaining the characteristics of the Nyquist plot.

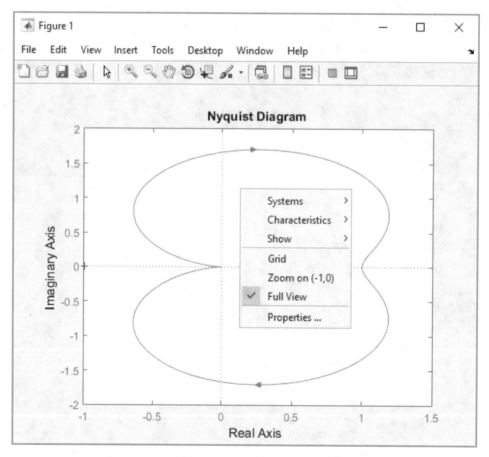

Figure 2.28: Click the grid in order to add the axes grids to the drawn Nyquist plot.

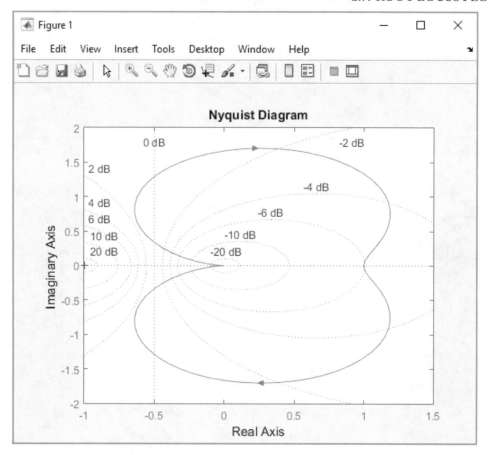

Figure 2.29: Nyquist plot of H with grid axes.

2.7   ROOT LOCUS PLOT

You can draw the root locus plot of a dynamical system with the aid of `rlocus` command (Fig. 2.30). For instance, you can draw the root locus of $H(s) = \frac{2s^2+5s+1}{s^2+2s+3}$ with the aid of following commands:

```
>> H=tf([2 5 1],[1 2 3]);
```

```
>> rlocus(H), grid on
```

Click the Data Cursor icon in order to add a cursor to the plot. Move the added cursor; the required gain is shown for each point (Fig. 2.31).

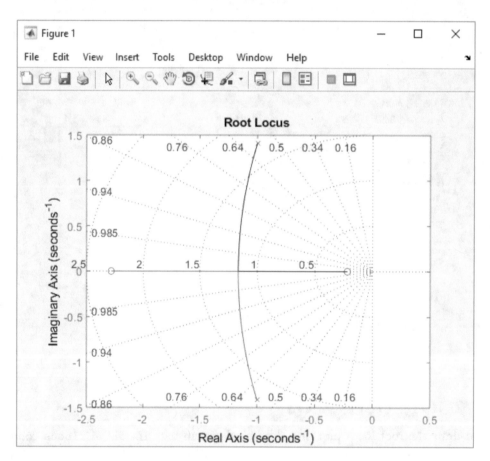

Figure 2.30: Root locus of $H(s) = \frac{2s^2+5s+1}{s^2+2s+3}$.

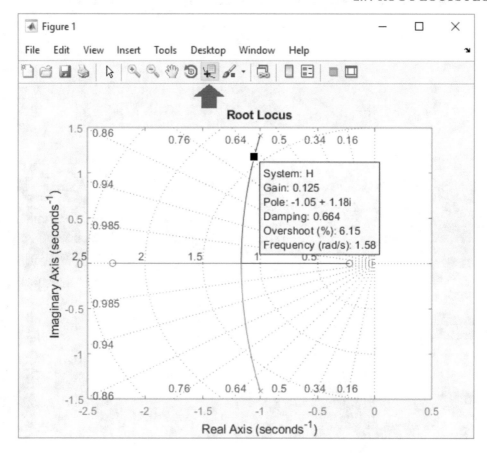

Figure 2.31: You can use the cursor to see the required gain. According to the shown box, gain of 0.125 places one of the closed loop poles at $-1.05 + 1.15i$. You can use the command pole(feedback(0.125*H,1) to check it (H=tf([2 5 1],[1 2 3])).

There is another command to draw the root locus plot: rltool. With the aid of rltool you can see the movement of all the poles at the same time. Type the following commands:

```
>> H=tf([2 5 1],[1 2 3]);
```

```
>> rltool(H), grid on
```

The window shown in Fig. 2.32 will appear.

Click the Root Locus Editor for Loop Transfer. The poles are shown with pink red points (Fig. 2.33).

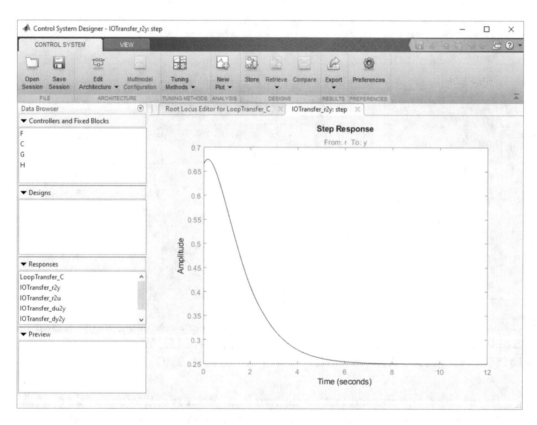

Figure 2.32: Control system designer window will appear after you run the `rltool` command.

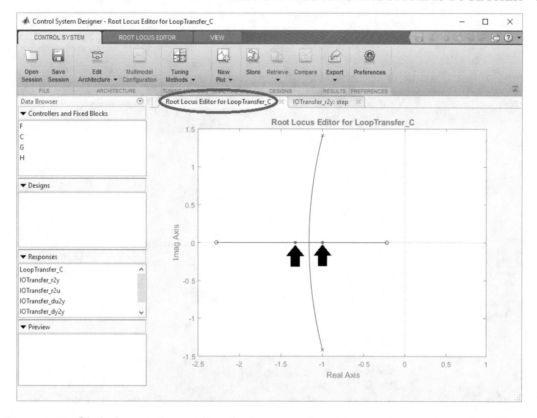

Figure 2.33: Click the root locus editor for loop transfer tab to see the root locus plot. Closed loop poles are shown with black arrows.

Move the closed-loop poles to desired location. To do this, left-click the pink points and move the points to the desired location without releasing the left mouse button. You can see the movement of all points at the same time. Required gain and other useful information are shown in the bottom of the window (Fig. 2.34).

2.8   CONNECTING THE SYSTEMS TOGETHER

Consider a block diagram such as the one shown in Fig. 2.35.

The closed-loop transfer function can be calculated with the aid of following commands:

```
>>C=tf([1.13 0.45],[1 0]);

>>P=tf([1],[1 3 3 1]);
```

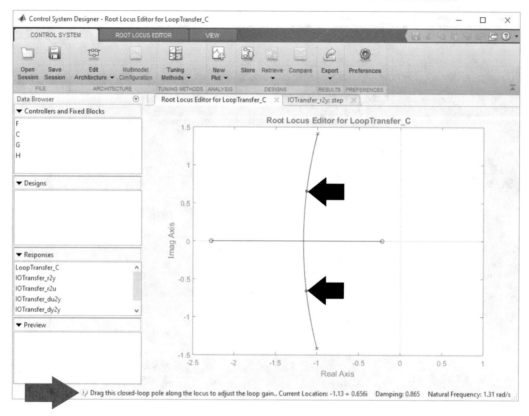

Figure 2.34: Moving the poles to desired location.

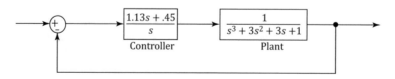

Figure 2.35: Sample feedback system.

```
>>CL=feedback(C*P,1)
```

You can see the step response of closed-loop system with the >>step(feedback(C*P,1)) command.

Consider a block diagram such as the one shown in Fig. 2.36.

In this case, the closed-loop transfer function can be calculated as follows:

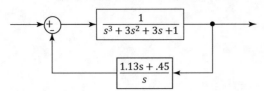

Figure 2.36: Sample feedback system.

```
>>C=tf([1.13 0.45],[1 0]);

>>P=tf([1],[1 3 3 1]);

>>CL=feedback(P,C)
```

Consider the block diagram shown in Fig. 2.37 (the feedback is positive in this block diagram).

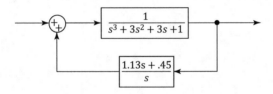

Figure 2.37: Sample feedback system. Note that feedback is positive.

In this case, the closed-loop transfer function can be calculated as follows:

```
>>C=tf([1.13 0.45],[1 0]);

>>P=tf([1],[1 3 3 1]);

>>CL=feedback(P,C,+1)
```

Consider the connection shown in Fig. 2.38.

The transfer function of connection shown in Fig. 2.38 can be obtained with the aid of series command, i.e., >> series(C,P).

The transfer function of connection shown in Fig. 2.39 can be obtained with the aid of parallel command, i.e., >>parallel(C,P).

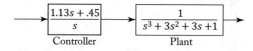

Figure 2.38: **Series connection of systems.**

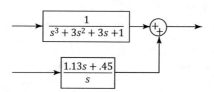

Figure 2.39: **Parallel connection of systems.**

CHAPTER 3

Introduction to Simulink®

3.1 INTRODUCTION

Simulink® is a graphical programming environment for modeling, simulating, and analyzing multidomain dynamical systems. Using Simulink®, the user can simulate the system behavior without using any code! The user only drag and drop the required blocks from the library to the simulation file and connect them. This chapter introduces this important tool with the Aid of some examples.

3.2 RUNNING THE Simulink®

In order to run the Simulink®, simply type the following command in the command window:

```
>>simulink
```

You can even click the Simulink® icon in the menu bar in order to run Simulink® (Fig. 3.1).

Figure 3.1: The Simulink® icon.

Select the Blank Model (Fig. 3.2).

Simulink® opens a new empty simulation file for you (Fig. 3.3).

Simulink® Library Browser contains the blocks which are used during the simulation. After the user finds the desired block inside the Simulink® Library Browser, drag and drop the block to the simulation file (Figs. 3.4 and 3.5).

Click the icon shown in Fig. 3.6 to open a new simulation file (if needed). You can press Ctrl+N as well.

A new simulation file will be opened (Fig. 3.7).

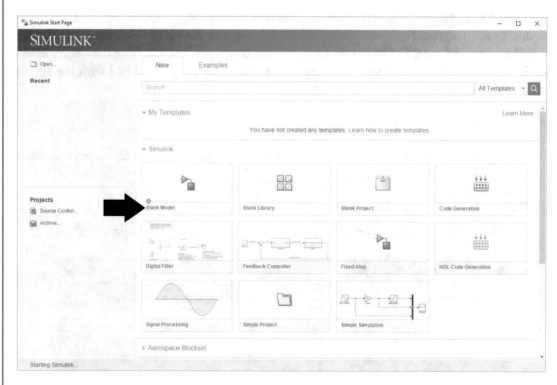

Figure 3.2: Simulink's first window.

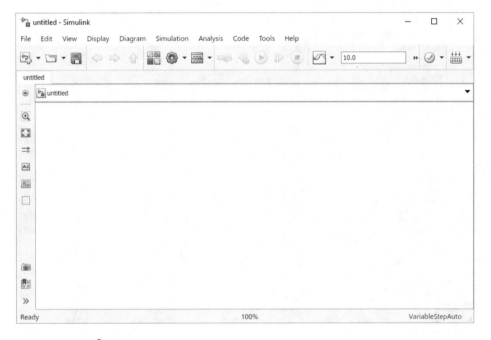

Figure 3.3: Simulink® environment.

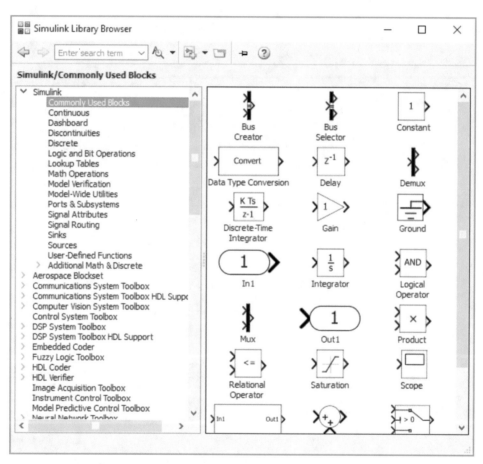

Figure 3.4: Simulink® library browser.

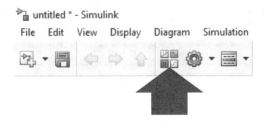

Figure 3.5: Use the shown icon if Simulink® doesn't show the library browser.

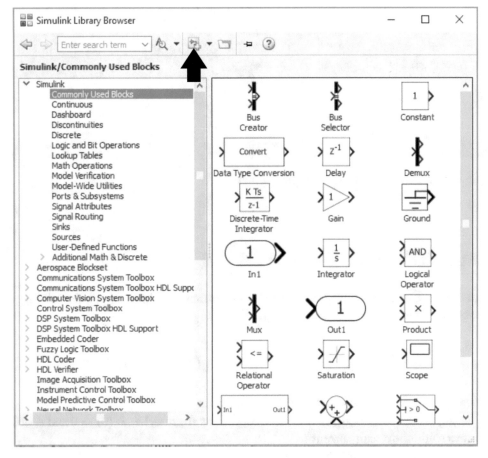

Figure 3.6: Opening a new simulation file.

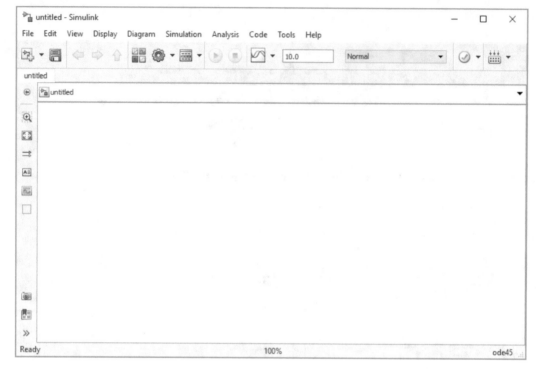

Figure 3.7: Simulation file. Blocks are placed here.

3.3   SEARCHING FOR BLOCKS

You can use the Enter search term box (Fig. 3.8) to search for a block if you don't know where it is. Just write what you look for and press the Enter key.

3.4   EXAMPLE 1: Simulink® OF A RLC CIRCUIT

Consider a simple RLC circuit such as the one shown in Fig. 3.9.

The key is closed at $t = 0$. All the initial conditions are zero, i.e., $i_L(0) = 0$ and $v_C(0) = 0$. We want to study the circuit current.

According to Kirchhoff voltage law,

$$L\frac{d^2i(t)}{dt^2} + R\frac{di(t)}{dt} + \frac{1}{C}i(t) = 0 \quad i(0) = 0 \frac{di(0)}{dt} = \frac{E}{L}. \tag{3.1}$$

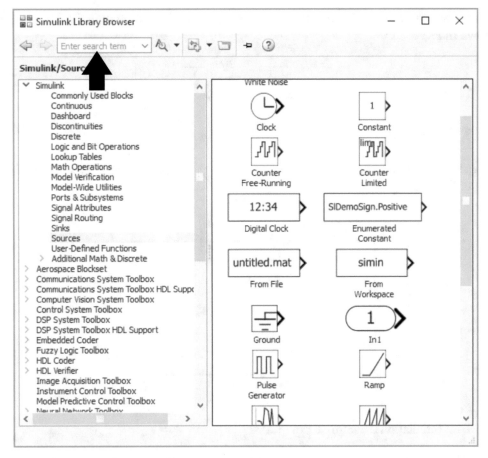

Figure 3.8: Use the Enter search term to find the desired blocks.

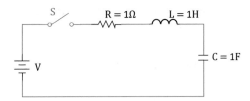

Figure 3.9: Simple RLC circuit.

We want to simulate this Differential Equation in Simulink®. To do this, we rearrange the obtained equation as

$$\frac{d^2 i(t)}{dt^2} = -\frac{R}{L}\frac{di(t)}{dt} - \frac{1}{LC}i(t) \quad i(0) = 0 \frac{di(0)}{dt} = \frac{E}{L}. \tag{3.2}$$

Use the Simulink® Library Browser to add two integrators to the simulation file. Connect them as shown in Fig. 3.10.

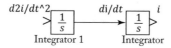

Figure 3.10: Two integrators are place in the simulation file.

Add two gain blocks, a summer and a scope to the simulation file, and connect them as shown in Fig. 3.11.

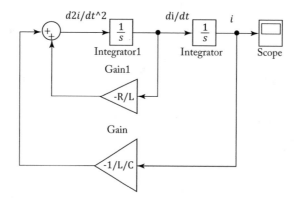

Figure 3.11: Simulation diagram of Eq. (3.2).

Double-click on the gain blocks and set their Gain box (see Figs. 3.11 and 3.12) to $-R/L$ and $-1/L/C$. We will define the value of R, L, and C later.

Simulink® has different solvers to solve the systems dynamic equation. The user can set his/her desired solver. To set the solver, click on the gearbox icon (Fig. 3.13).

Select the desired solver from Solver: drop down list (Fig. 3.14).

ode45 is a quite good solver for this example. Simulink® shows the active solver, i.e., the one which will be used during the simulation, in the bottom right side (Fig. 3.15).

We filled the values of gain blocks with $-R/L$ and $-1/L/C$. So, we must define the used variables before running the simulation. Otherwise the simulation will fail and an error message will be produced.

To define the values of aforementioned variables, go to command prompt and enter the commands shown in Fig. 3.16.

The differential equation has the $i(0) = 0$, $\frac{di(0)}{dt} = \frac{E}{L}$ set of initial conditions. To set the initial conditions, double-click the integrators and set the Initial condition box (Fig. 3.17). Set the initial condition of left integrator to E/L and the right one to 0 (Fig. 3.18).

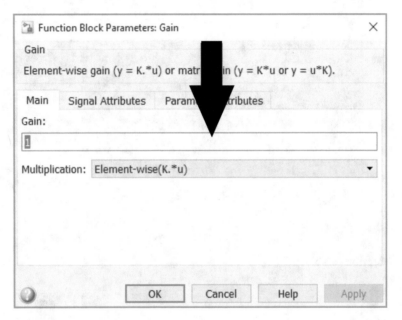

Figure 3.12: Setting the gain box.

Figure 3.13: Model configuration parameters icon.

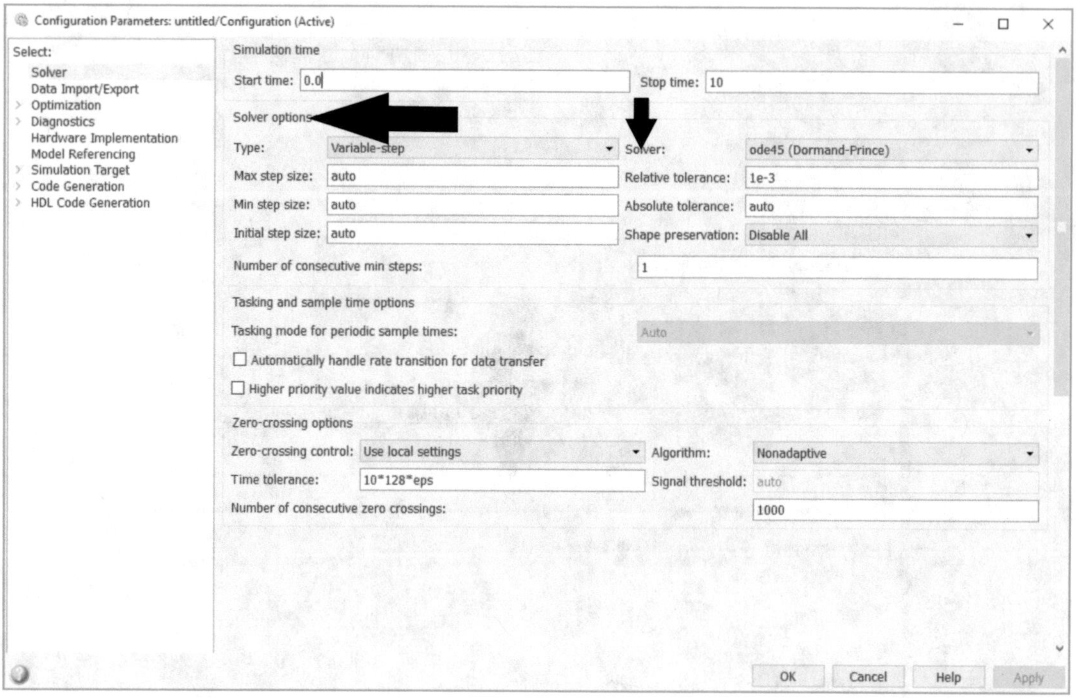

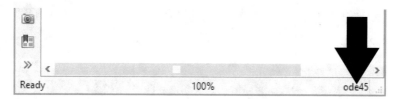

Figure 3.14: Selection of solver.

Figure 3.15: Active solver is shown in the right button of the screen.

Figure 3.16: Defining the R, L, C, and E variables.

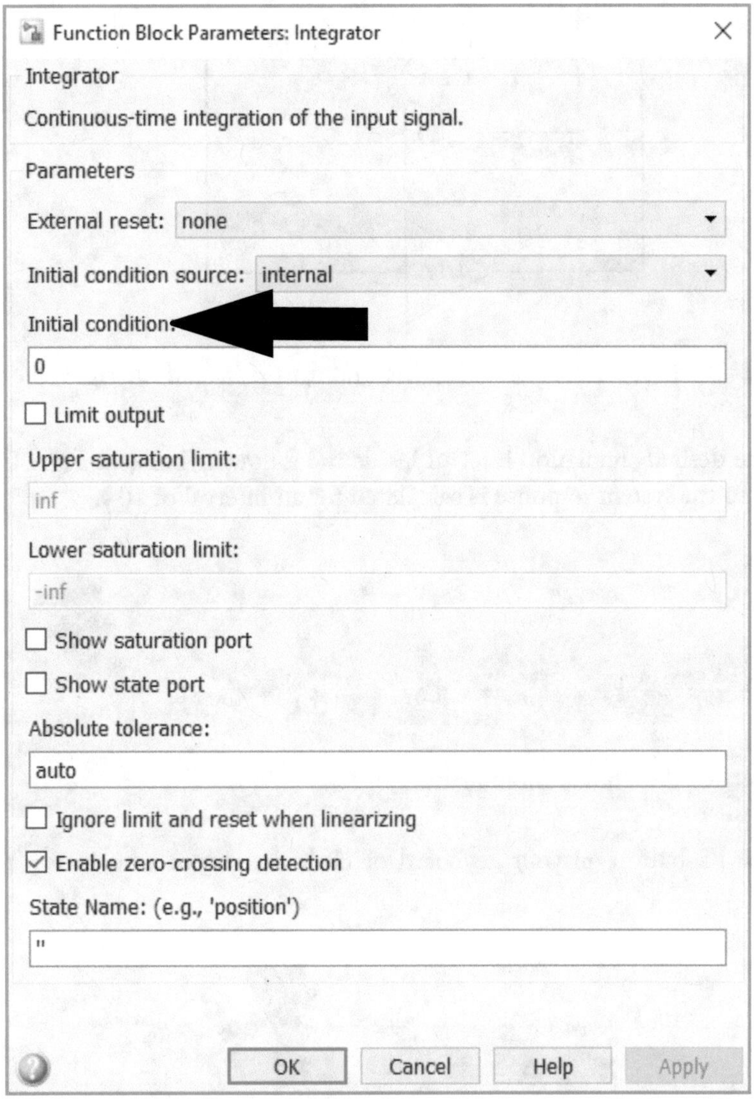

Figure 3.17: Setting the initial conditions.

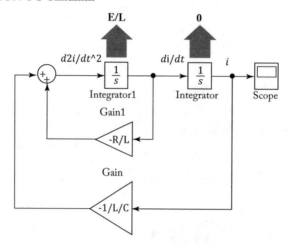

Figure 3.18: Set the initial condition of left integrator to E/L and the right one to 0.

Enter the desired simulation interval inside the following box (see Fig. 3.19). The box is filled with 10, so the system response is calculated for an interval of 10 s.

Figure 3.19: Setting the simulation interval.

Press the F5 button on your keyboard or click the shown icon to run the simulation (Fig. 3.20).

Figure 3.20: Run simulation icon.

After simulation is finished, double-click the scope block to see the current (Fig. 3.21).

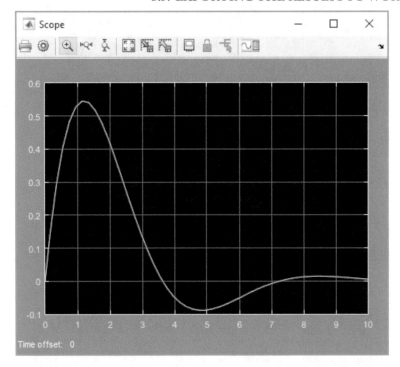

Figure 3.21: Simulation result (circuit current).

3.5   EXPORTING THE RESULTS TO WORKSPACE

You can export the simulation results to workspace for further process. To do this, add a To Workspace block to your simulation file and connect it to the rest of system as shown in Fig. 3.22.

Double-click the To Workspace block and set the Save format box to Structure With Time (Fig. 3.23).

Run the simulation and after the simulation is finished come back to MATLAB® environment. As shown in Fig. 3.24, two variables are added to workspace: simout and tout.

You can draw the plot of circuit current with the aid of following command. The result is shown in Fig. 3.25.

```
>>plot(simout.time,simout.signals.values(:),'k'),grid minor, hold on
```

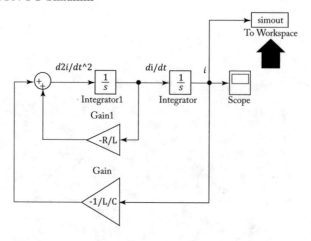

Figure 3.22: Addition of to workspace block to the simulation file.

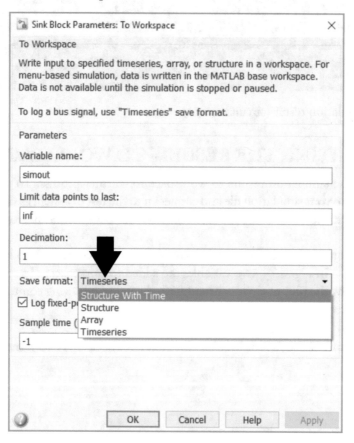

Figure 3.23: Selection of structure with time.

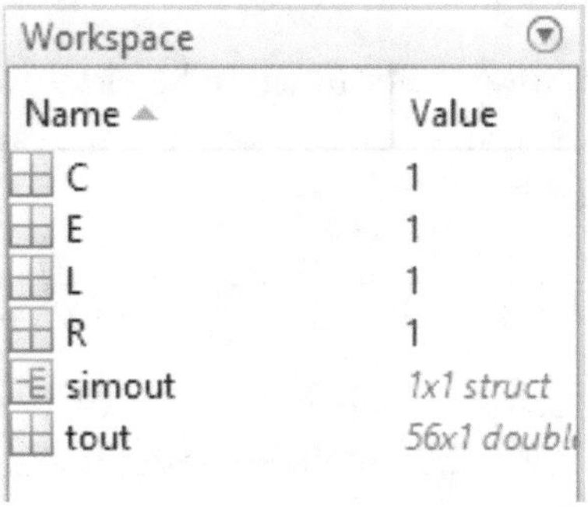

Figure 3.24: Simout and tout are added to the workspace variables.

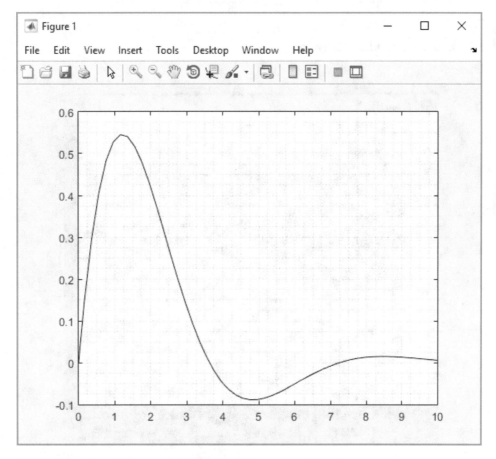

Figure 3.25: Drawing the imported data (inductor current).

3.6   EXAMPLE 2: SIMULATION IN TRANSFER FUNCTIONS

You can do the simulations contain transfer function blocks.

According to Fig. 3.9, the transfer function between the capacitor voltage and input source is:

$$H(s) = \frac{V_C(s)}{E(s)} = \frac{\frac{1}{Cs}}{R + Ls + \frac{1}{Cs}} = \frac{1}{LCs^2 + RCs + 1}. \tag{3.3}$$

We want to study the capacitor voltage in presence of a ramp voltage. We draw the simulation diagram shown in Fig. 3.26. The simulation result is shown in Fig. 3.27.

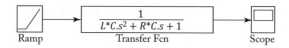

Figure 3.26: Simulation block diagram.

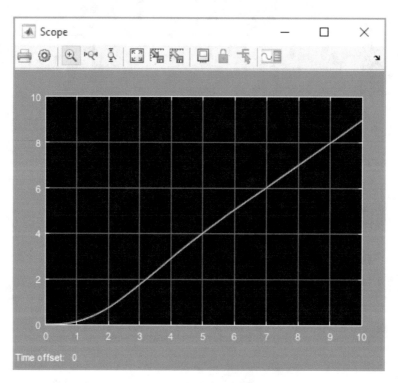

Figure 3.27: Simulation result.

3.7   EXAMPLE 3: SIMULATION OF ELECTRIC CIRCUITS WITH THE AID OF SIMSCAP LIBRARY

We give the system dynamics to Simulink® in the previous examples, i.e., we do the modeling phase by ourselves. We show how Simulink® can be used to simulate electric circuits in this example. Simulink® has a library named Simscape (see Fig. 3.28) which contains electrical blocks, i.e., resistors, capacitors, inductors, MOSFETs, AC/DC motors, etc. The Simscape library is used in this example.

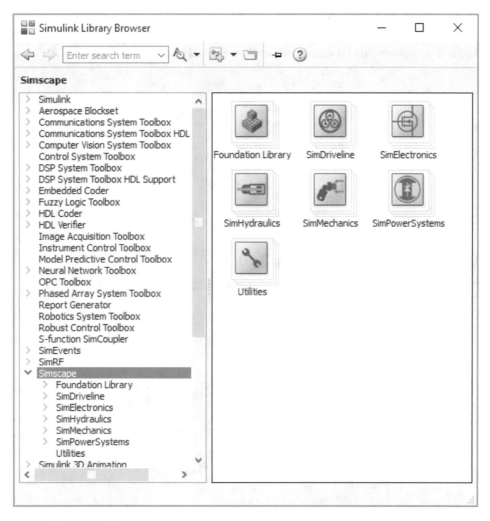

Figure 3.28: Simscape library.

We want to simulate the following RLC circuit and draw the graph of power drawn from DC voltage source (Fig. 3.29). The switch is closed at $t = 0$ and the system initial conditions are zero, i.e., $i_L(0) = 0$ and $v_C(0) = 0$.

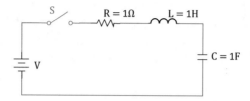

Figure 3.29: Simple RLC circuit.

Go to **Simscape>SimPowerSystems>Specialized Technology** (Fig. 3.30).

Figure 3.30: Specialized technology branch of SimPowerSystems.

Click the **Fundamental Blocks** (Fig. 3.31).

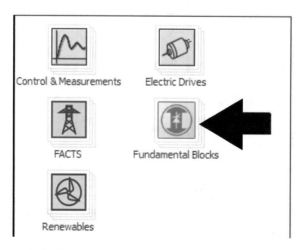

Figure 3.31: Fundamental blocks.

Select the **Elements** (Fig. 3.32).

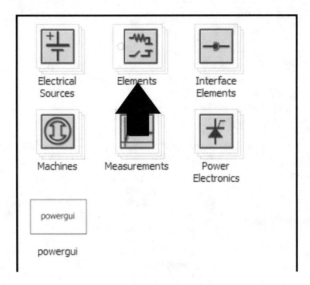

Figure 3.32: Elements branch of fundamental blocks.

Add three **Series RLC branch** (see Figs. 3.33 and 3.34) to the simulation file.

Double-click the Series RLC branch's. Use the Branch type (see Fig. 3.35) to convert the components to a resistor, an inductor, and a capacitor (see Fig. 3.36).

Double-click the components to set their values. Set the initial condition of inductor/capacitor as shown in Fig. 3.37.

Go to **Simscape>SimPowerSystems>Specialized Technology>Fundamental Blocks**. Click the **Electrical Sources** (see Fig. 3.38) and select a **DC Voltage Source** (see Fig. 3.39).

Drag and drop the DC source to simulation file (see Fig. 3.40).

We need voltmeter and ammeter to show the circuits voltage and currents. Voltmeter and ammeter can be found in **Simscape>SimPowerSystems>Specialized Technology>Fundamental Blocks>Measurements** (see Fig. 3.41).

Add voltmeter and ammeter to the circuit (see Figs. 3.42 and 3.43).

You can rotate a block place in the simulation file. To do this right-click the block and select the **Rotate and flip**. You can rotate a block by left-clicking it and then pressing the **Ctrl+R** (see Fig. 3.44).

All the simulations contains the SimPowerSystems blocks must contain a powergui block (Fig. 3.45). Otherwise the simulation can't be run and lead to an error message (see Fig. 3.46).

The complete simulation diagram is shown in Fig. 3.47

After running the simulation, the result shown in Fig. 3.48 is obtained.

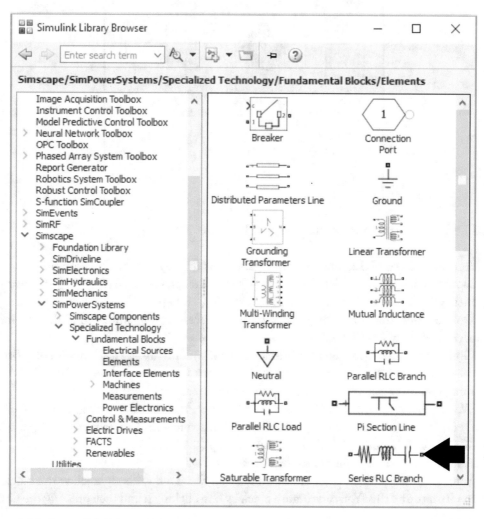

Figure 3.33: Series RLC branch.

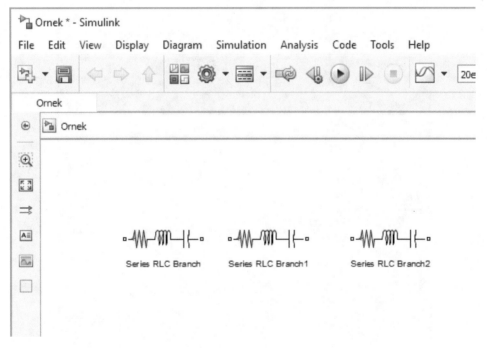

Figure 3.34: Addition of three series RLC branch to the simulation file.

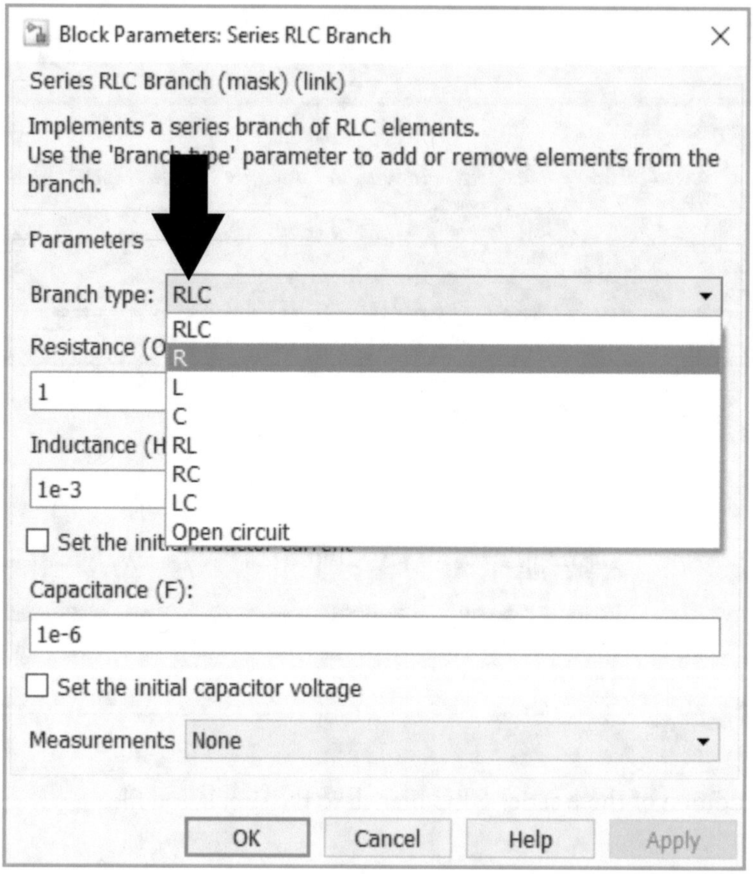

Figure 3.35: Different available options for branch type.

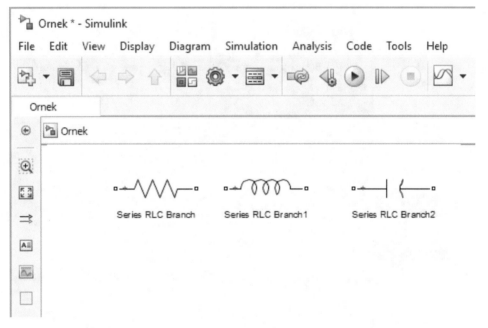

Figure 3.36: Conversion of series RLC branches to resistor, inductor, and capacitor.

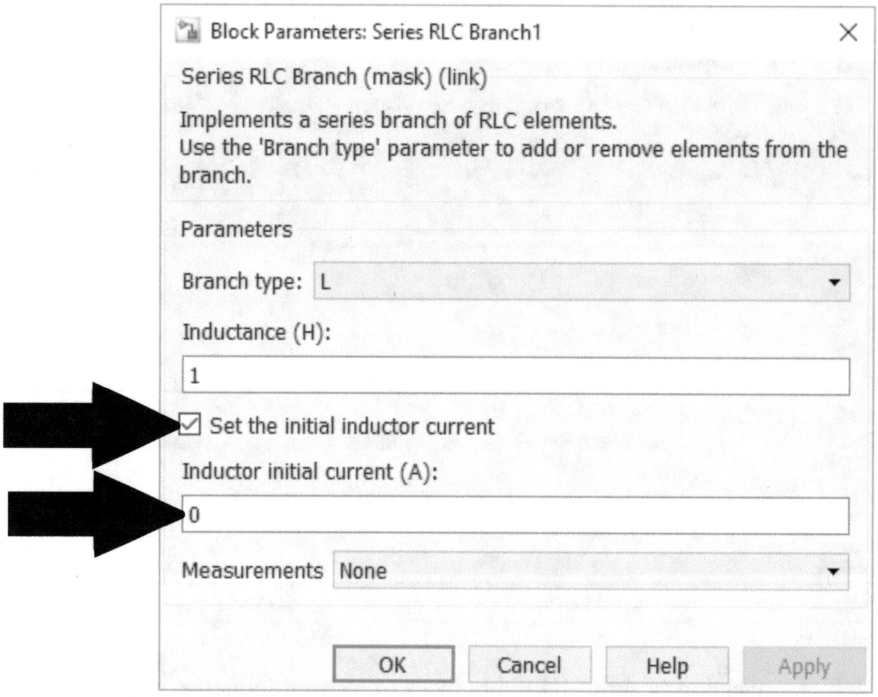

Figure 3.37: Setting the initial value of the inductor current.

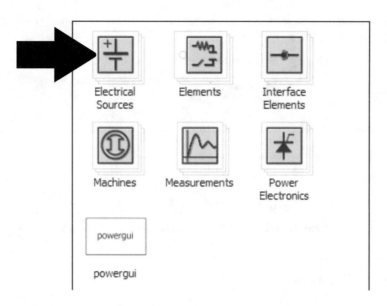

Figure 3.38: Electrical sources' branch of fundamental blocks.

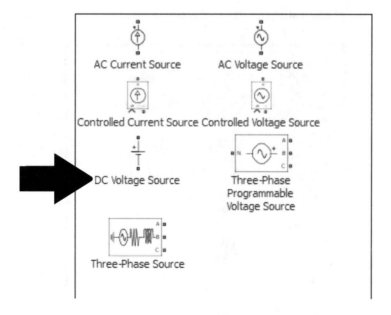

Figure 3.39: DC voltage source.

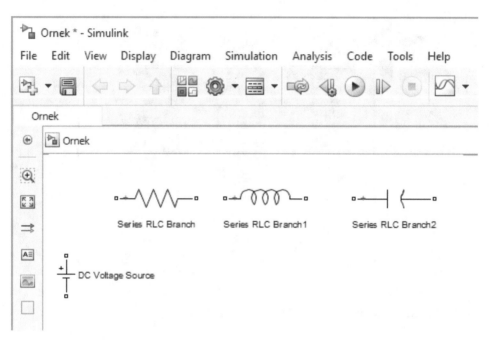

Figure 3.40: Addition of DC voltage source to the simulation file.

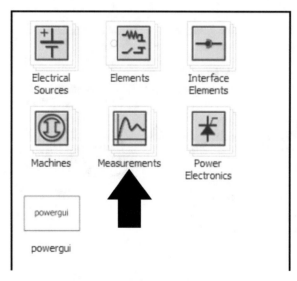

Figure 3.41: Measurements branch of fundamental blocks.

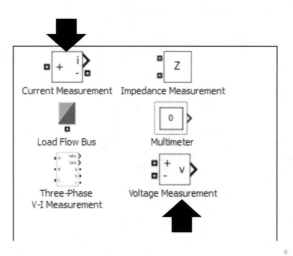

Figure 3.42: Ammeter and voltmeter blocks.

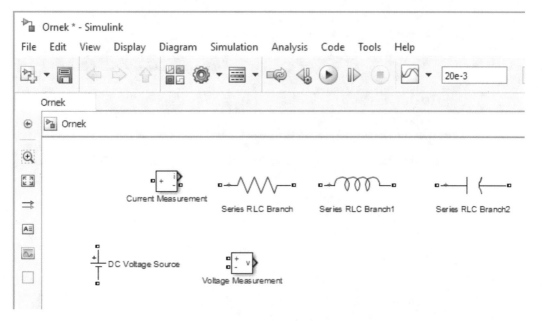

Figure 3.43: Addition of ammeter and voltmeter to the simulation file.

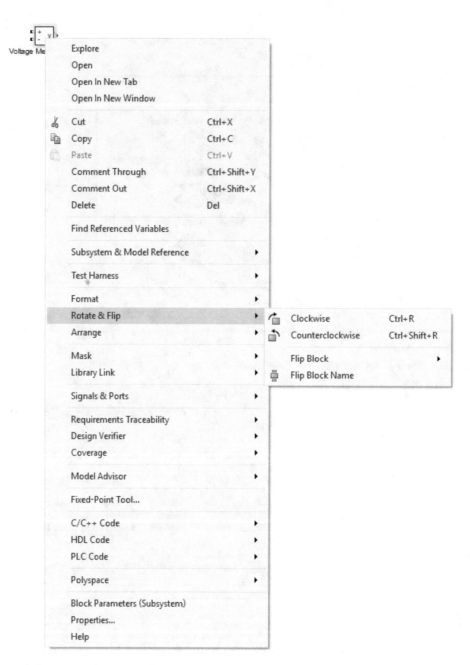

Figure 3.44: **Rotate and flip options.**

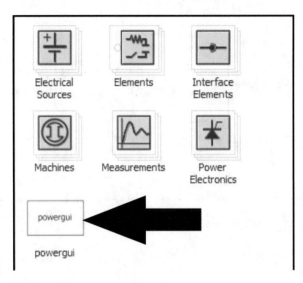

Figure 3.45: **Powergui block.**

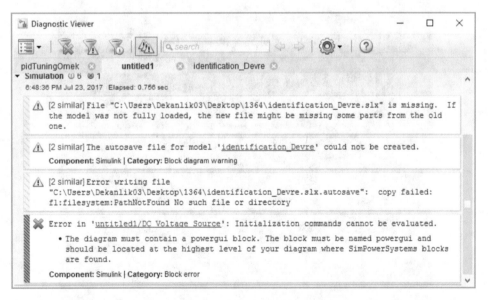

Figure 3.46: **An error message occurs when you don't add the powergui to your simulation file.**

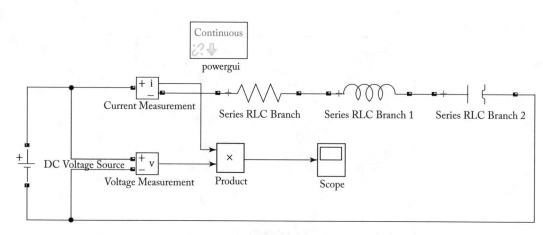

Figure 3.47: Addition of powergui to the simulation file.

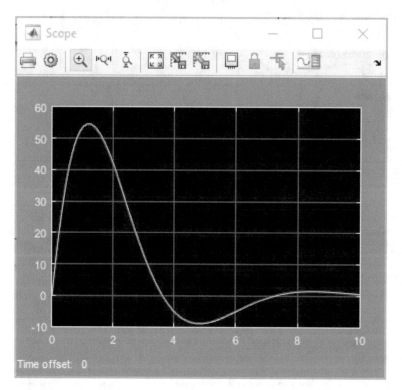

Figure 3.48: Simulation result (DC source power).

3.8   EXAMPLE 4: SIMULATION OF NONLINEAR SYSTEMS

We simulate a nonlinear system in this example. Consider the following system:

$$\begin{cases} \frac{d}{dt}i_L = \frac{v_{in}}{L} - (1-u)\frac{v_c}{L} \\ \frac{d}{dt}v_C = -\frac{1}{RC}v_c + (1-u)\frac{i_L}{C} \end{cases} \tag{3.4}$$

i_L and v_C are system states and u is the system input. u only takes two values: 0 and 1, i.e., $u \in \{0, 1\}$. All the initial conditions are zero, i.e., $i_L(0) = 0$ and $v_C(0) = 0$.

When $u = 1$,

$$\begin{cases} \frac{d}{dt}i_L = \frac{v_{in}}{L} \\ \frac{d}{dt}v_C = -\frac{1}{RC}v_c. \end{cases} \tag{3.5}$$

When $u = 0$,

$$\begin{cases} \frac{d}{dt}i_L = \frac{v_{in}-v_c}{L} \\ \frac{d}{dt}v_C = \frac{1}{C}i_L - \frac{1}{RC}v_c. \end{cases} \tag{3.6}$$

The simulation diagram of the system is shown in Fig. 3.49.

We want to stimulate the system with the periodic signal that is shown in Fig. 3.50. Such a signal is produced with the aid of Pulse Generator block.

Figure 3.51 shows the simulation results.

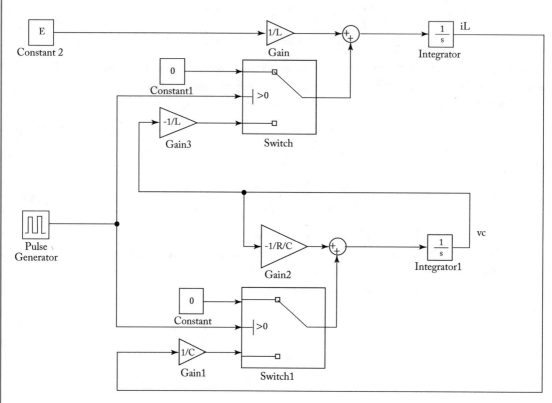

Figure 3.49: Simulation diagram of Example 4 in Section 3.8.

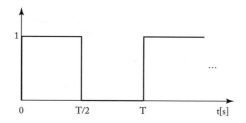

Figure 3.50: Excitation signal $\left(T = \frac{1}{25000} = 40\,\mu s\right)$.

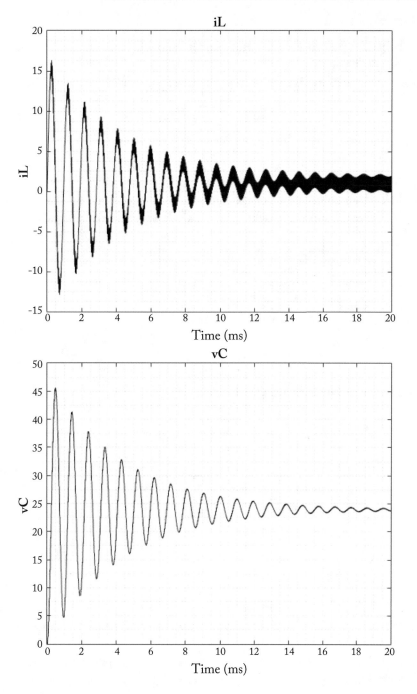

Figure 3.51: Simulation results.

3.9   SUBSYSTEM

Subsystems helps the user to produce more understandable simulation diagrams. It hides the unnecessary details inside it.

Consider a simulation diagram such as the one shown in Fig. 3.52.

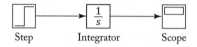

Figure 3.52: A simple simulation diagram.

Left-click on an empty point of the simulation file, don't release the left mouse button and draw a rectangle around the blocks. Simulink® draws a blue rectangle around the blocks (see Fig. 3.53).

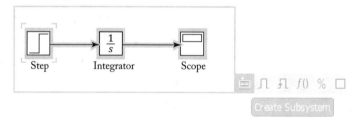

Figure 3.53: Draw a rectangle around the blocks.

Click the Create Subsystem (see Fig. 3.54).

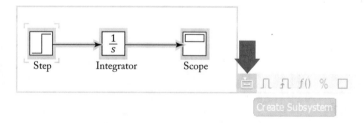

Figure 3.54: Create subsystem icon.

After you clicked the Create Subsystem icon, Simulink® makes the subsystem. Click on the word "Subsystem" to change it. Use descriptive names which shows the purpose of subsystem (see Fig. 3.55).

Double-click the subsystem. Simulink® opens the subsystem for you. Remove the Step and Scope blocks (click on the block and press the Delete key on keyboard) and replace them with In/Out blocks. In/Out blocks can be found in the Commonly Used Blocks tab (see Fig. 3.56).

Subsystem

Figure 3.55: Simulink® convert the selected blocks into a subsystem once create subsystem is clicked.

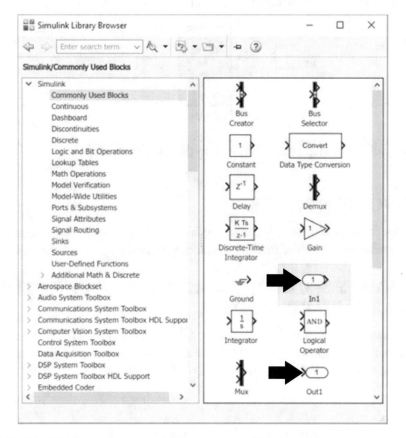

Figure 3.56: In/Out blocks.

The subsystems looks like the one shown in Fig. 3.57. It has input/output port. You can connect the subsystem to rest of system using it input/output ports.

Subsystem

Figure 3.57: Simulink® adds the input/output port to the subsystem.

Connect a Step and Scope blocks to the subsystems, as shown in Fig. 3.58.

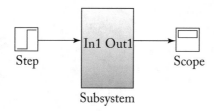

Figure 3.58: Addition of step and scope block to the simulation file.

3.10 FUNCTION BLOCK

You can use the function block to produce more understandable simulation diagrams. Study the following example to see how the function block can be used in your simulations.

3.11 EXAMPLE 5: USE OF FUNCTION BLOCK IN THE Simulink® SIMULATIONS

Consider a dynamical system with the equation $\frac{dy(t)}{dt} = 3 \times y(t) + x(t)^2$. $x(t)$ and $y(t)$ show the system input and output, respectively. We want to simulate the system with a sinusoidal input, $x(t) = \sin(t)$. The initial condition is zero ($y(0) = 0$).

Solution:
1st way:
You can draw the following simulation diagram (see Fig. 3.59).

2nd way:
You can use the function block to simulate the system. Function block can be found at SIMULINK>User Defined Functions>Fcn (see Fig. 3.60).

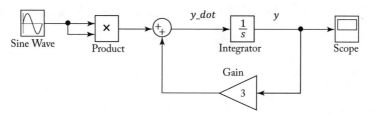

Figure 3.59: Simulink® diagram of Example 5 in Section 3.11. No function block is used in this simulation.

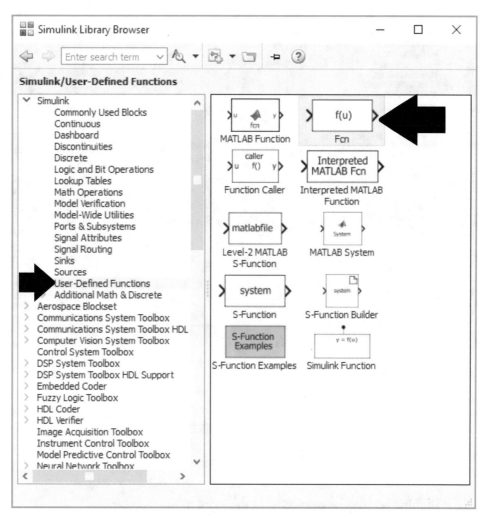

Figure 3.60: Function block.

Add a Mux block (see Fig. 3.61) to the simulation file and connect it to the Fcn block (Fig. 3.62).

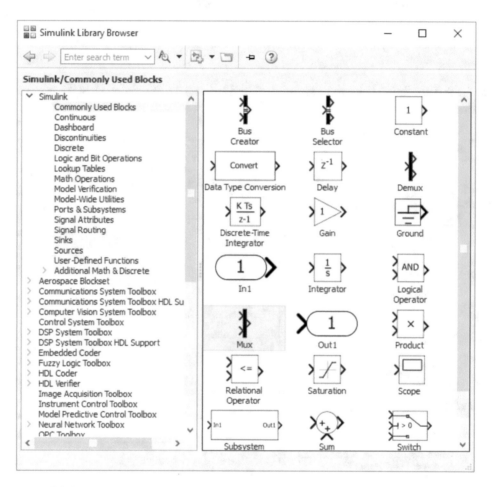

Figure 3.61: Multiplexer block.

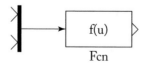

Figure 3.62: Multiplexer block is connected to the input of function block.

Double-click the Fcn block and fill the Expression box, as shown in Fig. 3.63.

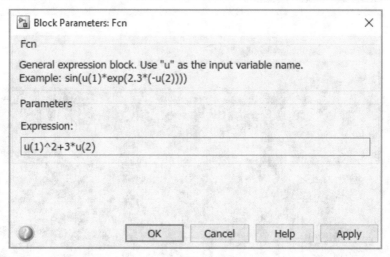

Figure 3.63: Filling the expression box of function block.

The $u(1)$ and $u(2)$ are the inputs of the multiplexer block. The $u(1)$ is the topmost input (see Fig. 3.64).

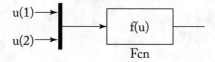

Figure 3.64: Input of function block are $u(1)$ and $u(2)$. $u(1)$ is the topmost input.

Figure 3.65 shows the simulation diagram of the system.

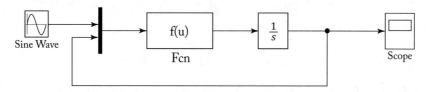

Figure 3.65: Simulink® diagram of Example 5 in Section 3.11. Function block is used in this simulation.

Use of function blocks (when possible) is suggested. The simulation result is shown in Fig. 3.66.

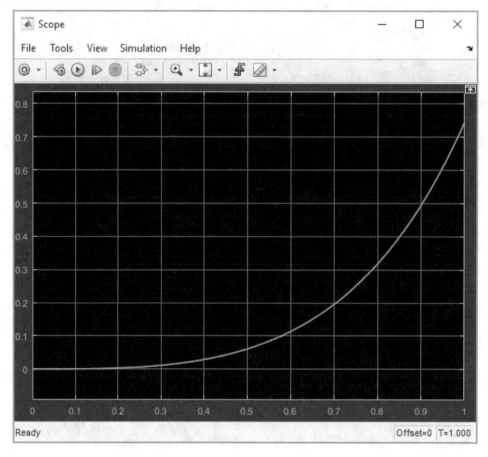

Figure 3.66: **Simulation result.**

3.12   SIMULATION OF DISCRETE TIME SYSTEMS

We want to simulate a discrete time system with the following equation:

$$2 \times x(n-1) + 6 \times x(n-2) = x(n), \quad x(-1) = 11 \, x(-2) = 7.$$

We want to obtain $x(n)$, $n \geq 0$. Some of the first terms can be calculated manually

$$x(0) = 2 \times 11 + 6 \times 7 = 64$$
$$x(1) = 2 \times 64 + 6 \times 11 = 194$$
$$x(2) = 2 \times 194 + 6 \times 64 = 772.$$

We use the **Delay** block (Fig. 3.67) to simulate the system.
Add two delay blocks to your simulation and connect them together, as shown in Fig. 3.68.

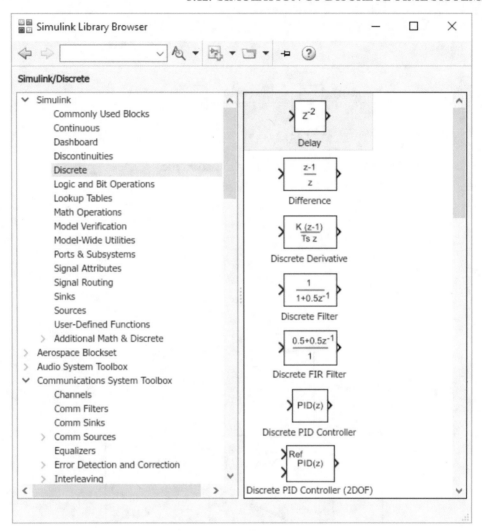

Figure 3.67: **Delay block.**

The system can be simulated with the aid of simulation diagram shown in Fig. 3.69.

The initial condition can be set with the aid of Initial condition box. Double-click the delay blocks (Figs. 3.70 and 3.71), after the block is opened; set their corresponding initial condition value (Fig. 3.72).

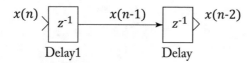

Figure 3.68: Two delay blocks are added to the simulation file.

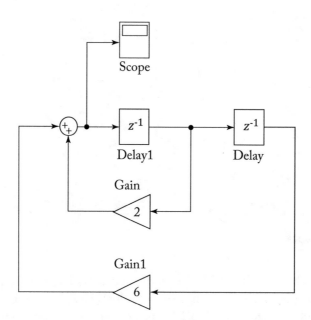

Figure 3.69: Completed simulation file.

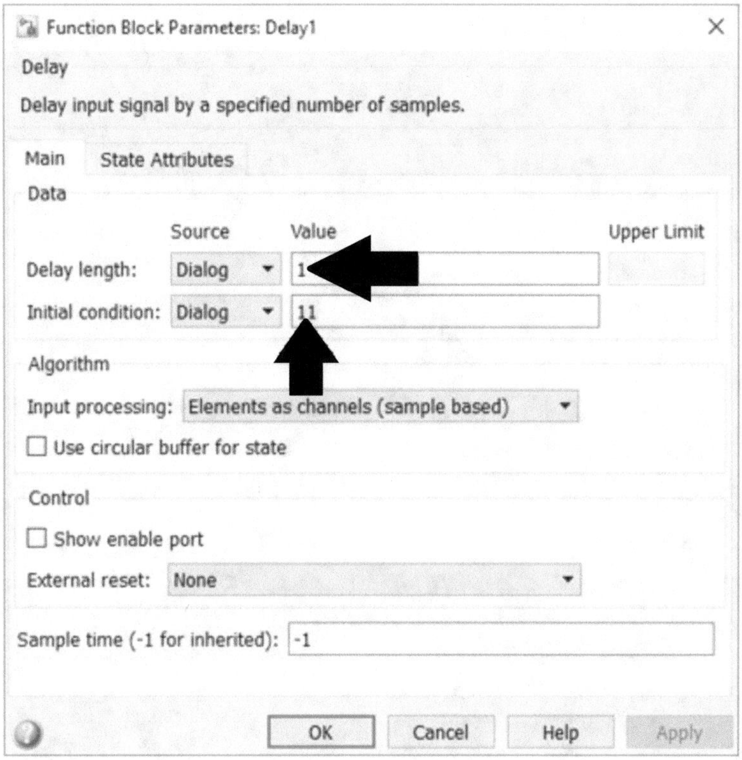

Figure 3.70: Setting the delay1.

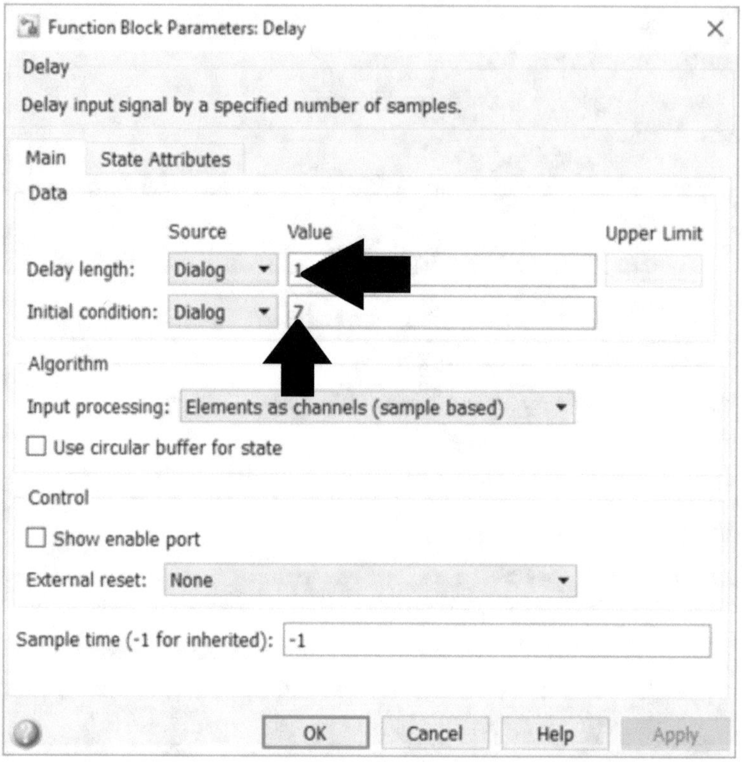

Figure 3.71: Setting the delay.

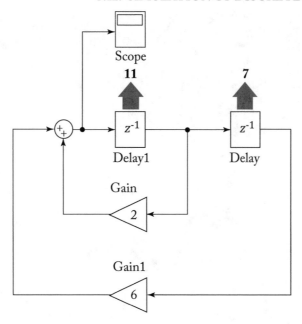

Figure 3.72: Initial value of left delay block is set to 11 and the right block is set to 7.

Click the gear icon and set the Solver to Discrete (no continuous state). See Figs. 3.73 and 3.74.

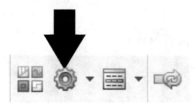

Figure 3.73: Model configuration parameters icon.

Simulation result is shown in Fig. 3.75.

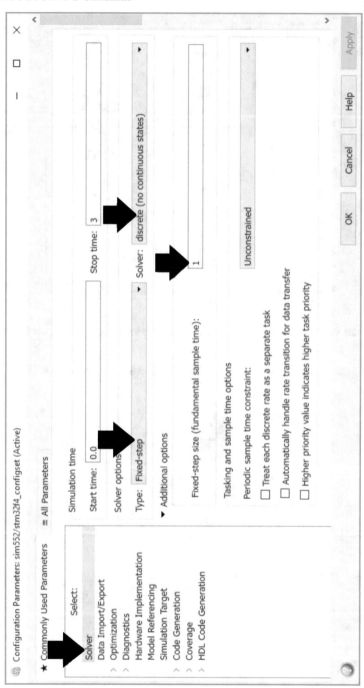

Figure 3.74: Setting the solver type to discrete (no continuous states).

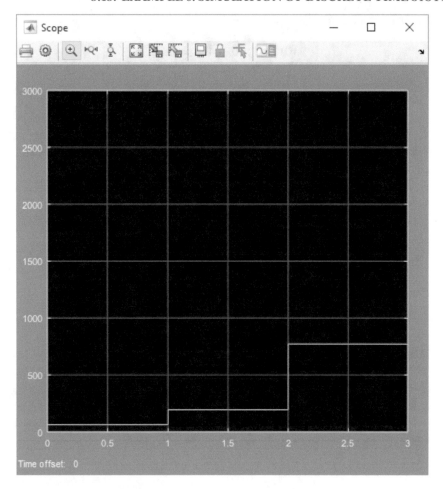

Figure 3.75: Simulation result.

3.13 EXAMPLE 6: SIMULATION OF DISCRETE TIME SYSTEMS

We want to simulate another system. Consider the following system:

$$y(n) - 3 \times y(n-1) = 4^{n-1}, \ y(-1) = 0.$$

We write the system as:

$$y(n) = 3 \times y(n-1) + 4^{n-1}, \ y(-1) = 0.$$

The first few terms can be calculated:

$$y(0) = 3 \times 0 + 4^{-1} = 0.25,$$

$$y(1) = \frac{3}{4} + 1 = \frac{7}{4} = 1.75,$$

$$y(2) = \frac{21}{4} + 4 = \frac{37}{4} = 9.25.$$

The system can be simulated with the aid of following simulation diagram (see Fig. 3.76).

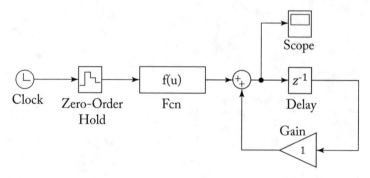

Figure 3.76: Simulation diagram of Example 6 in Section 3.13.

In order to produce the excitation term (4^{n-1} term), we used a Clock block, a Zero-Order Hold block, and a function block (Fig. 3.77). The clock block produces a continuous output. Zero-Order Hold converts the output of the clock block to a staircase signal (see Figs. 3.78, 3.79, and 3.80).

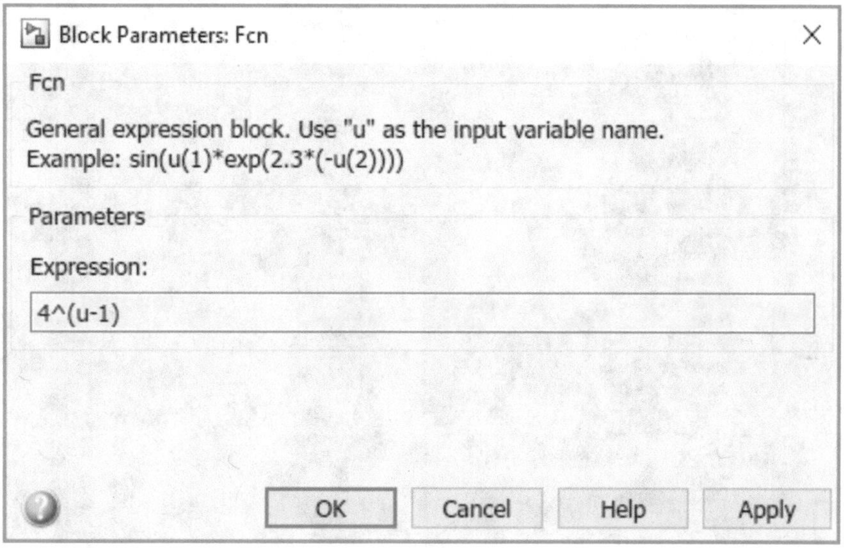

Figure 3.77: Setting the function block.

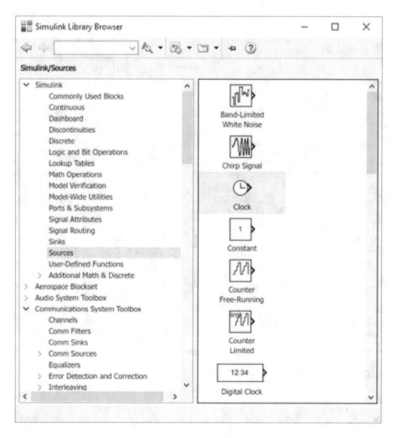

Figure 3.78: The clock block.

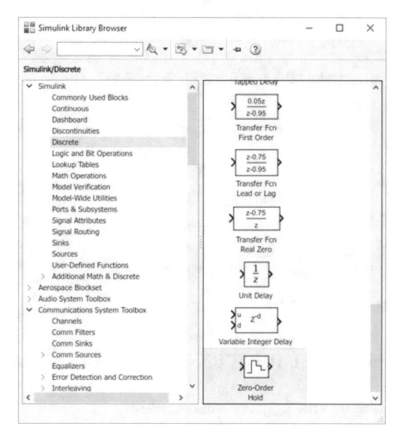

Figure 3.79: The zero-order hold block.

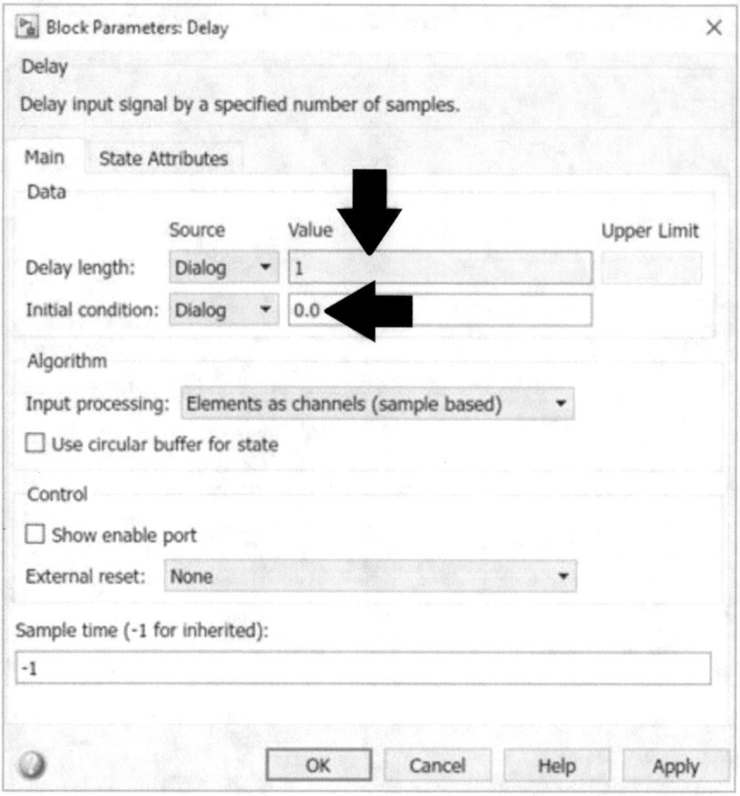

Figure 3.80: The delay block setting.

3.14   EXAMPLE 7: SIMULATION OF DISCRETE TIME SYSTEMS

We want to simulate the following system in this example:

$$y(n) - \frac{1}{2}y(n-1) = \frac{1}{4}x(n) + \frac{1}{4}x(n-1), \; y(-1) = 0,$$

$y(n)$ is the system output and $x(n)$ is the input. $x(n)$ is unit step function,

$$x(n) = \begin{cases} 0 & n < 0 \\ 1 & n \geq 0. \end{cases}$$

The given equation can be rearranged as:

$$y(n) = \frac{1}{2}y(n-1) + \frac{1}{4}x(n) + \frac{1}{4}x(n-1), \; y(-1) = 0.$$

The first few terms can be calculated as:

$$y(0) = \frac{1}{4} = 0.25,$$

$$y(1) = \frac{5}{8} = 0.625,$$

$$y(2) = \frac{13}{16} = 0.812,$$

$$y(3) = \frac{29}{32} = 0.906.$$

The simulation diagram of the system is shown in Fig. 3.81. The used blocks settings are given in Figs. 3.82–3.84.

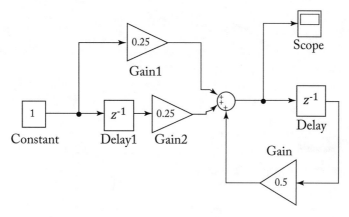

Figure 3.81: The simulation diagram of Example 7 in Section 3.14.

3.15 SHOWING TWO OR MORE SIGNALS SIMULTANEOUSLY

You can show two or more signals simultaneously with the aid of Mux block (see Fig. 3.85).

In the following simulation file, three different signals are shown on an Scope. Simulation result is shown in Fig. 3.86. Multiplexer settings are shown in Fig. 3.87 and the simulation result is shown in Fig. 3.88.

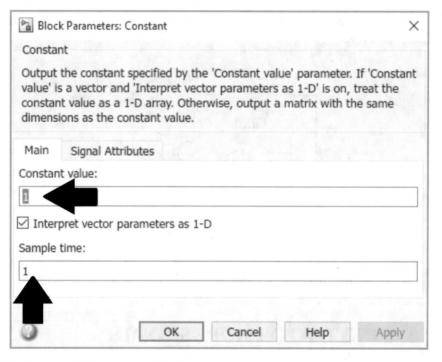

Figure 3.82: Setting of the constant block.

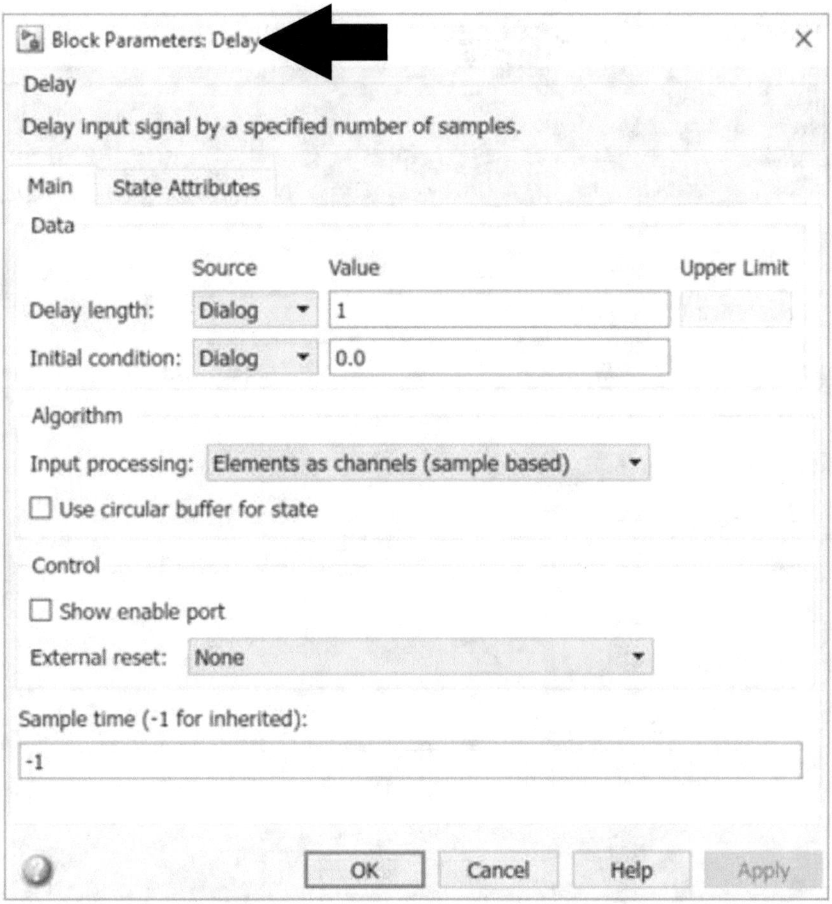

Figure 3.83: Setting of the delay block.

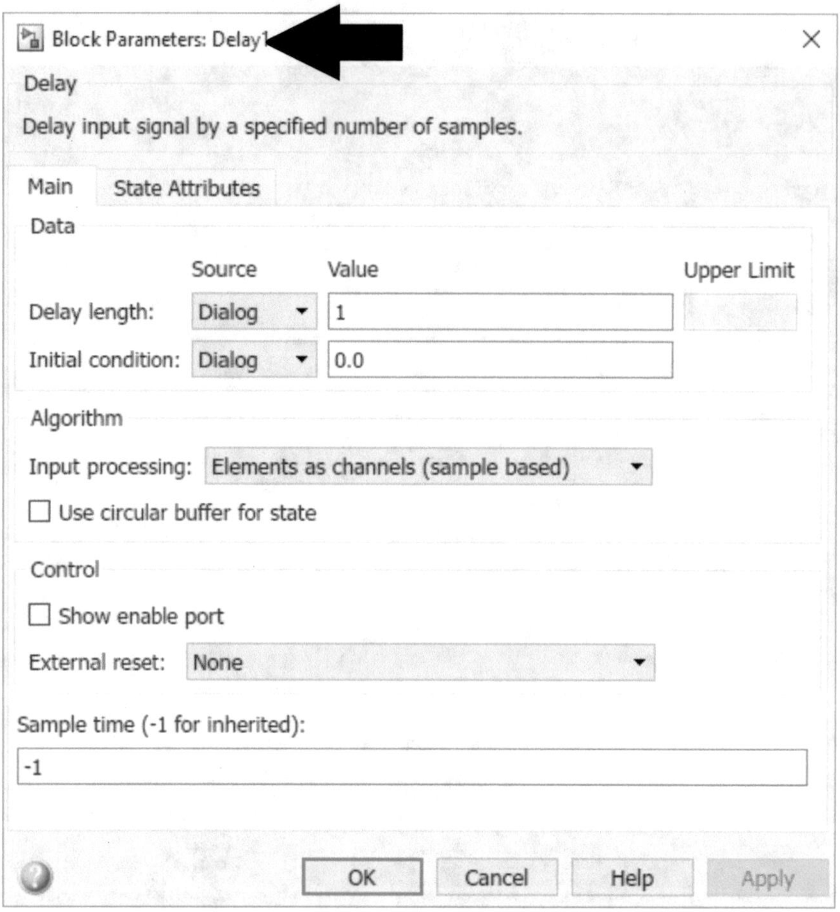

Figure 3.84: Setting of the delay1 block.

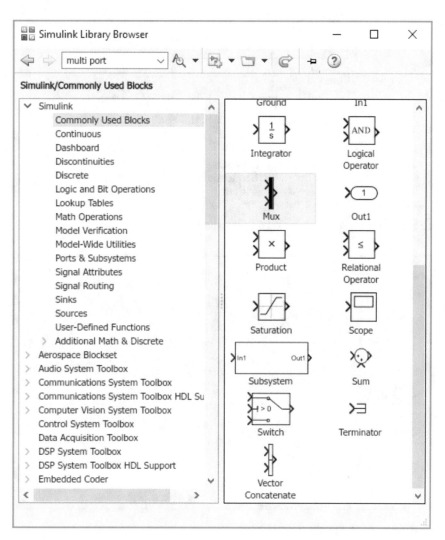

Figure 3.85: **Multiplexer block.**

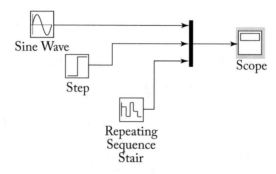

Figure 3.86: Showing three different signals on the same scope.

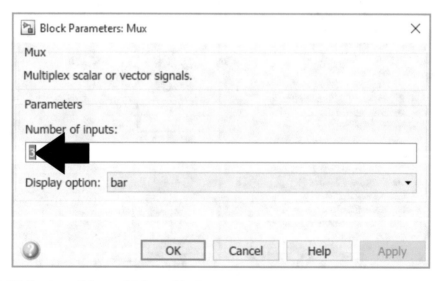

Figure 3.87: Setting of the multiplexer block.

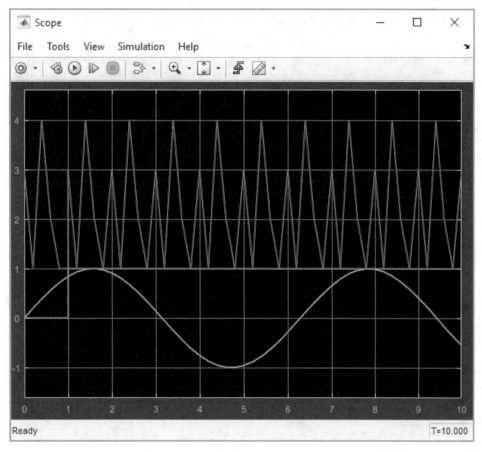

Figure 3.88: Simulation result.

3.16 SIMULATION OF A CLOSED-LOOP CONTROL SYSTEM IN Simulink®

We want to simulate the closed-loop control system shown in Fig. 3.89.

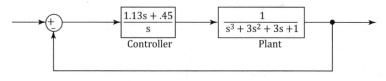

Figure 3.89: Sample closed-loop system.

Add two Transfer Function block to the simulation file (Figs. 3.90 and 3.91).

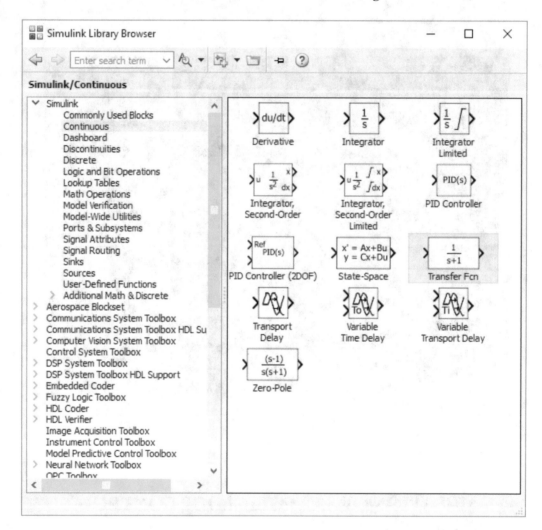

Figure 3.90: Transfer function block.

Figure 3.91: Adding two transfer function block to the simulation file.

Add a summer to the simulation file (Figs. 3.92 and 3.93).

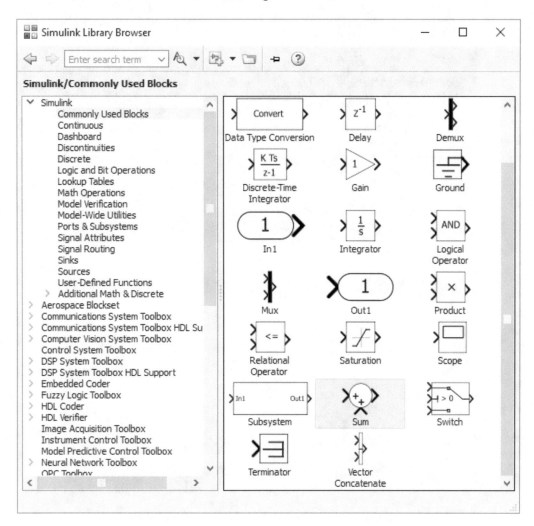

Figure 3.92: The summer block.

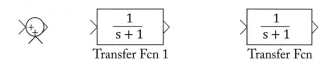

Transfer Fcn 1   Transfer Fcn

Figure 3.93: Addition of summer block to the simulation file.

Double-click on the summer and set the list of signs to $+-$ (Figs. 3.94 and 3.95).

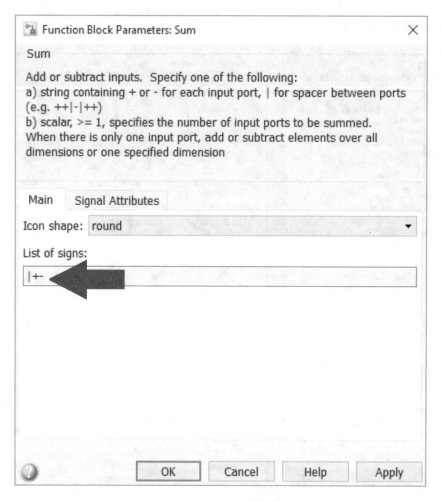

Figure 3.94: Set the List of signs to $+-$ in order to simulate negative feedback.

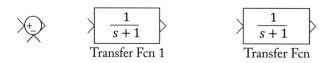

Figure 3.95: The summer can simulate negative feedback now.

Connect the blocks together (Fig. 3.96). Click on the "Transfer FC1" and "Transfer Fcn" labels and change them to "Controller" and "Plant," respectively. This have no effect on the simulation but it makes the simulation diagram more meaningful from the user point of view.

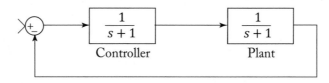

Figure 3.96: Connecting the blocks together.

Double-click on the Controller and Plant and set the transfer functions, as shown in Figs. 3.97 and 3.98. The simulation diagram looks like the one shown in Fig. 3.99 after you change the coefficients.

You can make the Plant a little bit bigger to see the equation of transfer function inside it (Fig. 3.100).

Add a step block to the simulation file (Figs. 3.101 and 3.102).

Double-click the simulation file. Step time is set to 1 by default (Fig. 3.103). This cause the block to produce the signal shown in Fig. 3.104. Turn the Step time to 0 if you want to stimulate the system with the one shown in Fig. 3.105.

Add a Scope block to the system (Fig. 3.106). An scope block is added to see the plant output (Fig. 3.107). Run the simulation with the aid of the button shown (Fig. 3.108). See the results in Fig. 3.109.

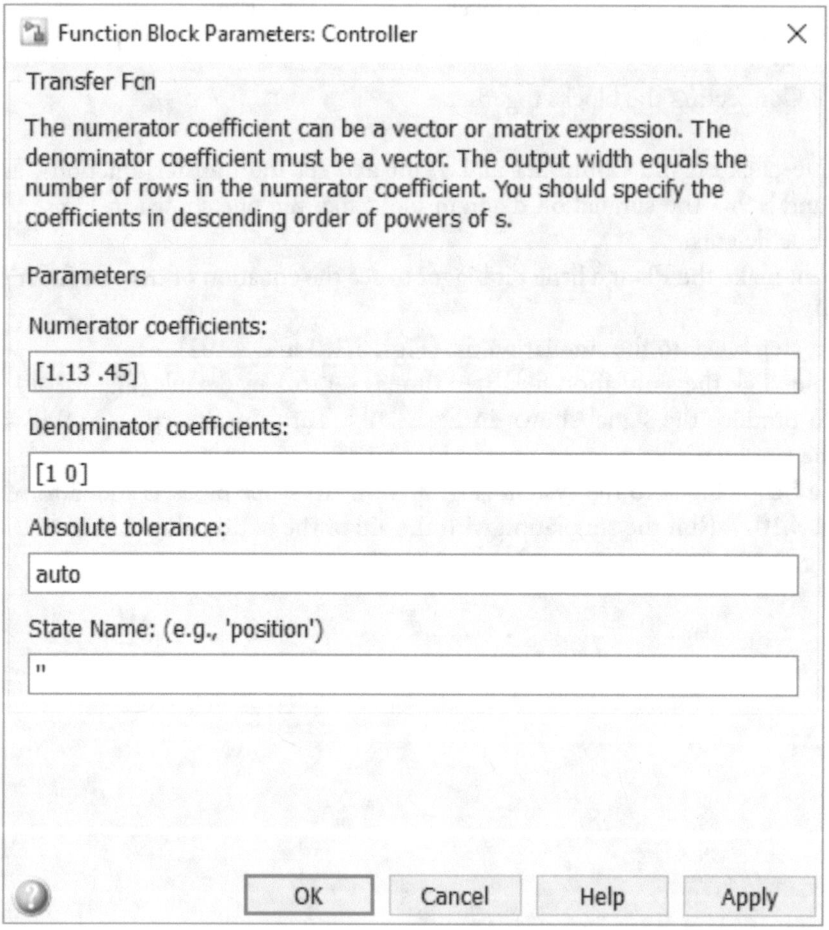

Figure 3.97: Entering the coefficients of controller.

Function Block Parameters: Plant   ✕

Transfer Fcn

The numerator coefficient can be a vector or matrix expression. The denominator coefficient must be a vector. The output width equals the number of rows in the numerator coefficient. You should specify the coefficients in descending order of powers of s.

Parameters

Numerator coefficients:

[1]

Denominator coefficients:

[1 3 3 1]

Absolute tolerance:

auto

State Name: (e.g., 'position')

"

| OK | Cancel | Help | Apply |

Figure 3.98: Entering the coefficients of plant.

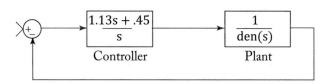

Figure 3.99: The simulation file after new coefficients entered.

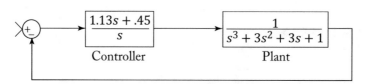

Figure 3.100: Drag the plant sides in order to show the transfer function inside it.

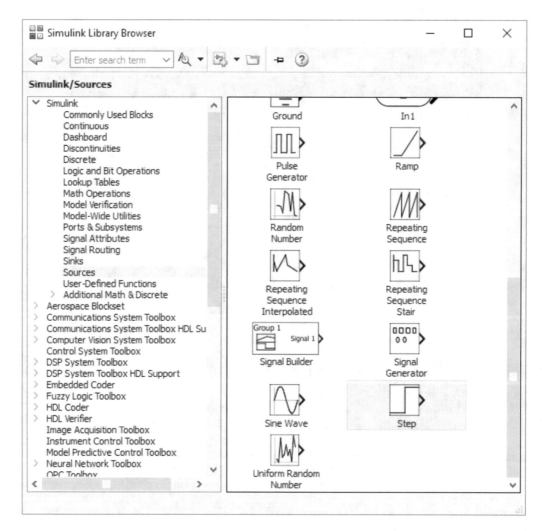

Figure 3.101: The step block.

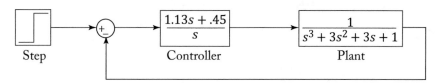

Figure 3.102: Completed simulation file.

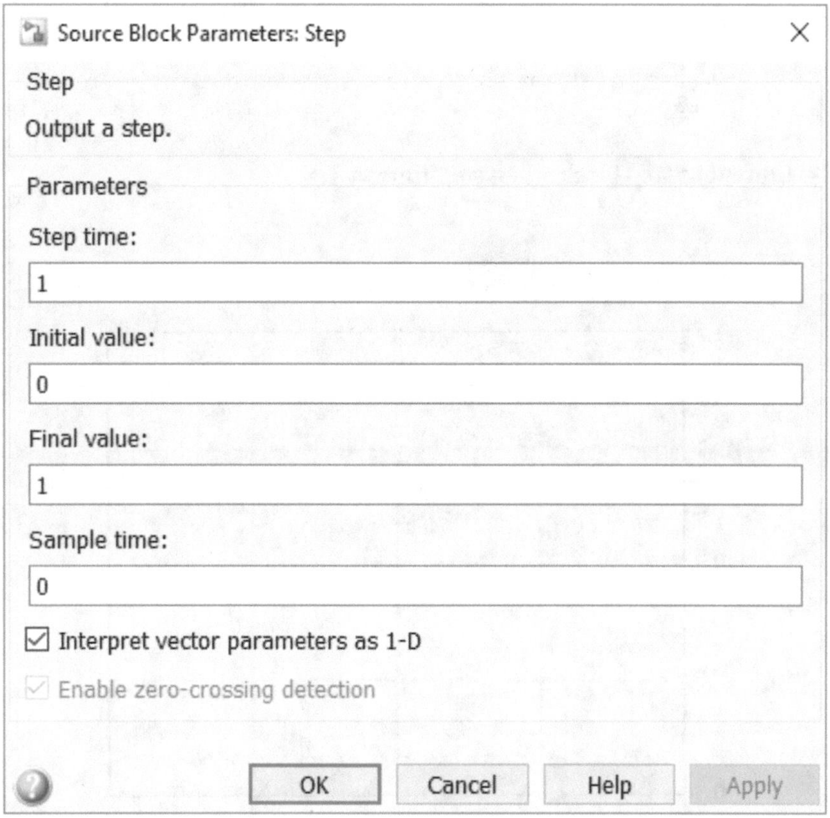

Figure 3.103: Step blocks setting.

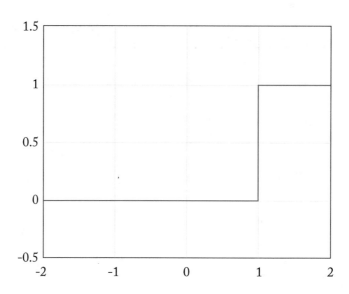

Figure 3.104: Output of step block when step time = 1.

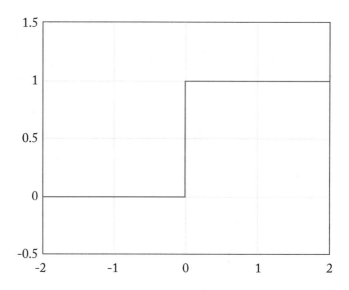

Figure 3.105: Output of step block when step time = 0.

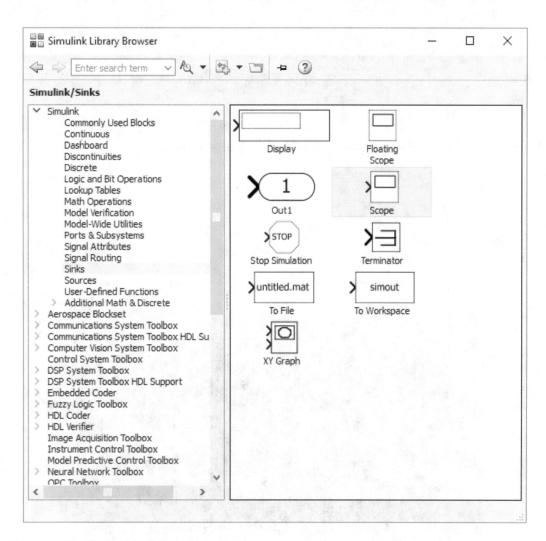

Figure 3.106: Scope block.

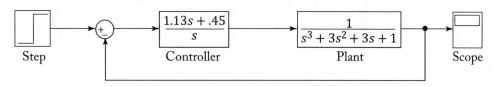

Figure 3.107: An scope block is added to see the plant output.

Figure 3.108: Run the simulation with the aid of the button shown.

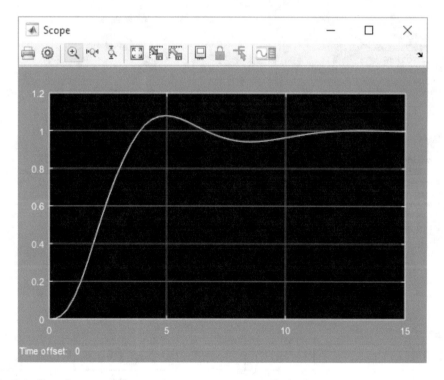

Figure 3.109: Simulation result.

CHAPTER 4

Controller Design in MATLAB®

4.1 INTRODUCTION

MATLAB® can be used to design different types of controllers. This chapter introduces some of the most important tools provided by MATLAB® to design linear controllers.

PID controllers are the most commonly used controller used in real world. About 90% of controllers used in industry are PID controllers. Fortunatly, MATLAB® has powerful functions to tune this important controller. Using MATLAB® a PID controller can be tuned within few minutes!

4.2 PID CONTROLLER DESIGN IN MATLAB®

MATLAB® has a powerfull function named `pidTuner` to design PID controllers. Its syntax is:

pidTuner(plantName)

We show how to use it, with the aid of an example.

We want to design a PID controller for

$$H(s) = \frac{s+1}{(s+2)(s+3)(s+4)} = \frac{s+1}{s^3 + 9s^2 + 26s + 24}.$$

We must define the dynamical equation in the MATLAB® environment. We do the job with the aid of one of the following commands.

```
>>H=tf([1 1],[1 9 26 24]);
```

or

```
>>H=zpk([-1],[-2 -3 -4],1);
```

The step function of this transfer function is shown in Fig. 4.1. So, there is a need to a controller, since it cannot track a step signal, i.e., the steady-state error is not zero. The systems is slow as well.

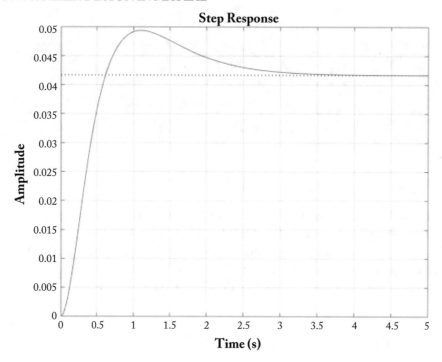

Figure 4.1: Step response of H.

We type the following command in MATLAB® command prompt:

```
>>pidTuner(H)
```

This opens the PID Tuner window (Fig. 4.2).

Select the controller type first. Available options are shown in Fig. 4.3. We select a PI for this application (we want to obtain zero steady-state error so we need the integrator term).

Use the sliders to obtain your desired response (Fig. 4.4).

You can do the tuning in the frequency domain if you prefer. If you like to do so, set the Domain to Frequency (Fig. 4.5).

When you switch to frequency domain, your slider changes to Bandwidth and Phase Margin, as shown in Fig. 4.6.

Change the sliders to obtain the desired response. The controller parameters, i.e., the PI controller coefficients are shown in the bottom right side (Fig. 4.7).

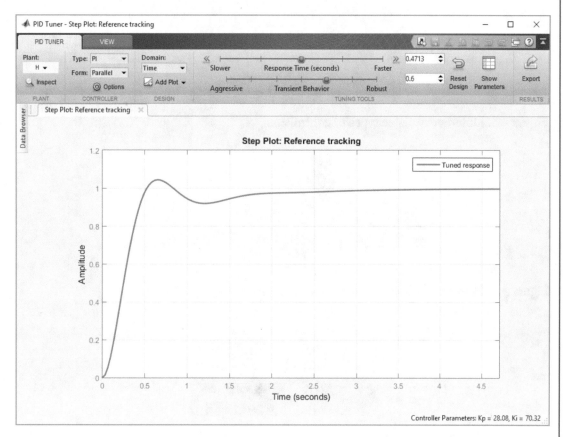

Figure 4.2: **PID tuner window.**

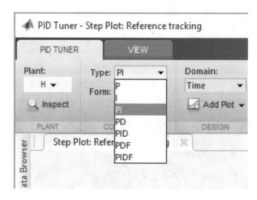

Figure 4.3: Different available types of controllers.

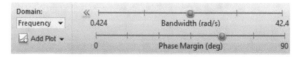

Figure 4.4: Move the sliders to tune the controller parameters.

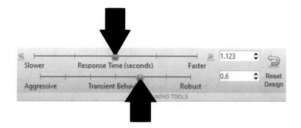

Figure 4.5: You can tune the controller in the frequency domain.

Figure 4.6: Tuning in the frequency domain.

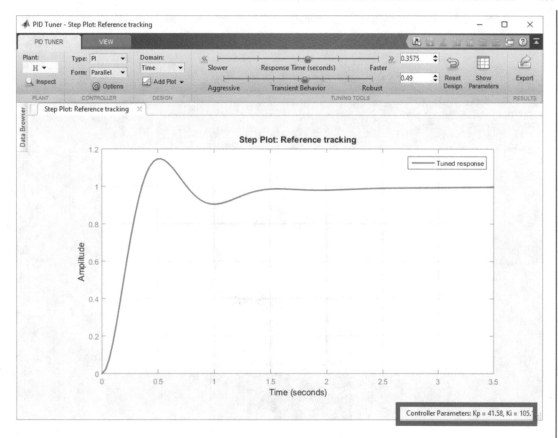

Figure 4.7: The controller parameters are shown in the bottom left side of the window.

Always ensure that the control signal (see Fig. 4.8) is within the desired range. Obtaining a desired output response (Fig. 4.7) is not enough; you must ensure the controller is able to produce the control signal as well. To ensure that control signals are withing their range, click the Add Plot icon and select the Controller Effort (Fig. 4.9).

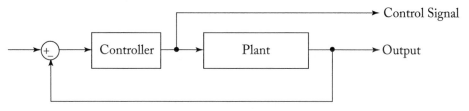

Figure 4.8: Control signal (control effort) is the signal produced by the controller.

The controller effort window will be added to the PID Tuner window (Fig. 4.10).

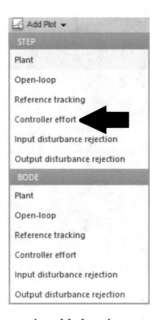

Figure 4.9: Plot of controller effort can be added to the main window by clicking the controller effort.

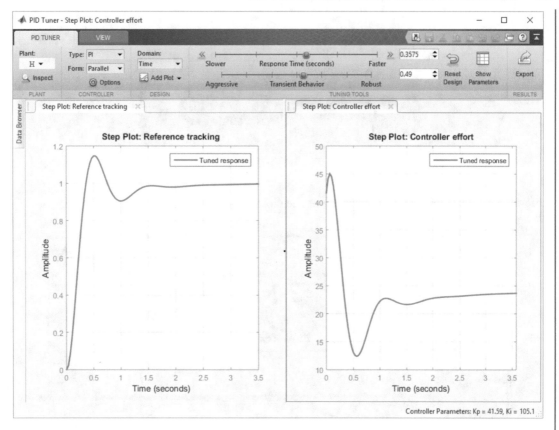

Figure 4.10: The designer can see the controller effort while he/she changes the sliders.

If the control signal is too large, repeat the design process, i.e., change the sliders until you obtain a more suitable response. Openning the controller effort plot is always suggested. Since the designer can see the control signal during the design process.

We want to regenerate the graphics shown in Fig. 4.10. Since the $Kp = 41.59$ and $Ki = 105.1$ the controller equation is $H_c(s) = 41.59 + \frac{105.1}{s}$. The following code (Fig. 4.11) draws the closed-loop step response. The result is shown in Fig. 4.12 and is the same with the graph shown in bottom left of the window (see Fig. 4.10).

The following code (Fig. 4.13) draws the control signal (control effort) for a step input. The result is shown in Fig. 4.14 and is the same with the graph shown in bottom right of the window (see Fig. 4.10).

Use the **Export** icon to export the designed controller into the MATLAB® workspace (Fig. 4.15).

```
Command Window                                    ⊙
    >> s=tf('s');
    >> H=tf([1 1],[1 9 26 24]);
    >> Hc=41.59+105.1/s;
    >> step(feedback(Hc*H,1)), grid on
fx >> |
```

Figure 4.11: Drawing the step response of the closed-loop system.

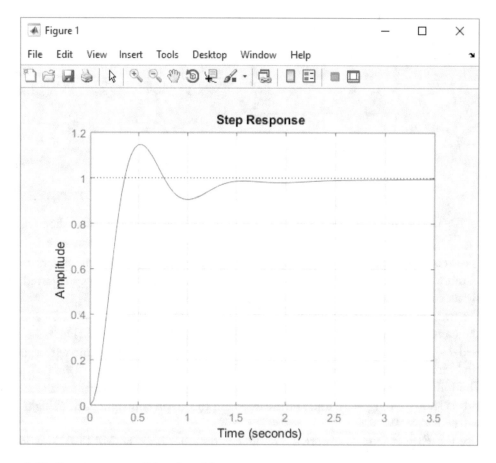

Figure 4.12: Step response of the closed-loop system.

```
Command Window                              ⊙
  >> s=tf('s');
  >> H=tf([1 1],[1 9 26 24]);
  >> Hc=41.59+105.1/s;
  >> step(Hc*feedback(1,Hc*H)), grid on
fx >> |
```

Figure 4.13: Drawing the control effort for a step reference.

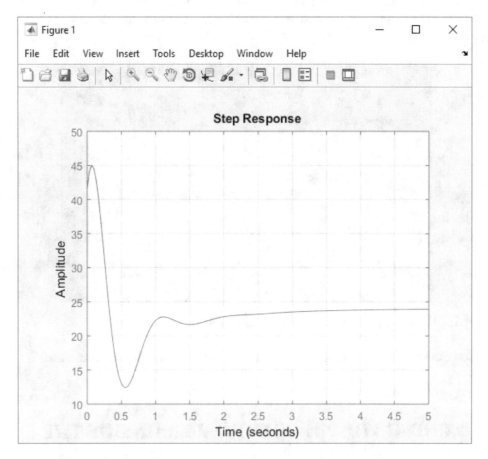

Figure 4.14: Control effort signal for a step reference.

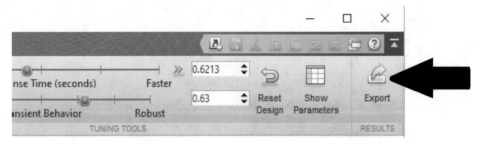

Figure 4.15: Export icon.

After you click the Export icon, the following window will appear (Fig. 4.16). Click OK to export the controller. The designed controller is saved in a variable named C inside the workspace.

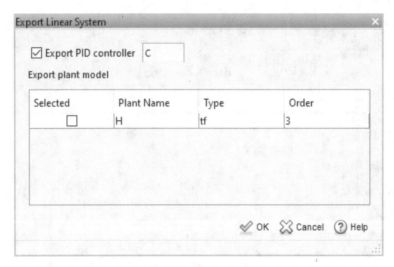

Figure 4.16: The designed controller is exported to Workspace.

You can see the closed-loop step response with the aid of following command:

```
>>step(feedback(H*C,1))
```

4.3 TUNING THE PID CONTROLLERS INSIDE THE Simulink® ENVIRONMENT

You can do the tuning of PID controller inside the Simulink® environment. For example, consider the system shown in Fig. 4.17.

Double-click the PID Controller block (Fig. 4.18).

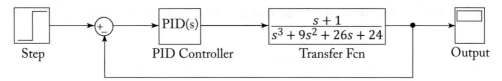

Figure 4.17: The Simulink® diagram of a closed-loop system.

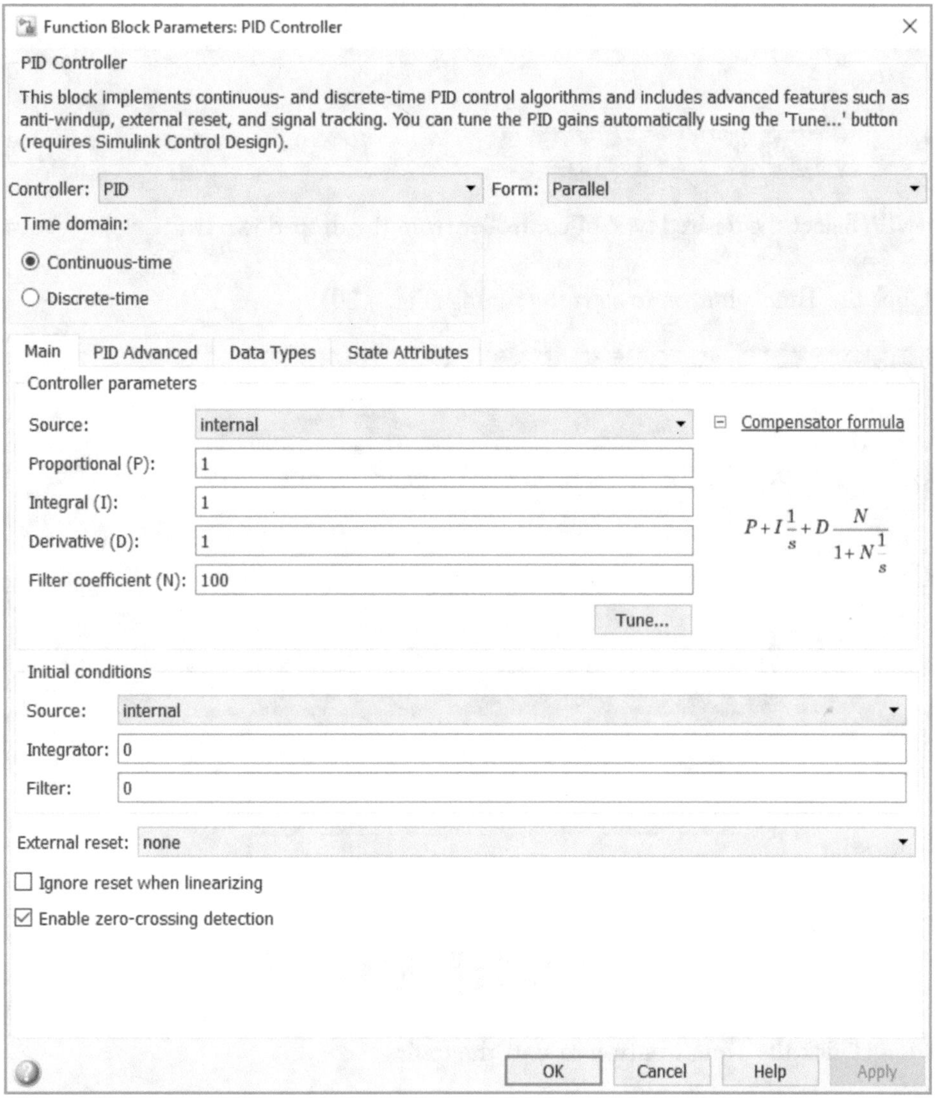

Figure 4.18: **PID Controller settings.**

Select your desired type of controller using the Controller drop-down list (Fig. 4.19).

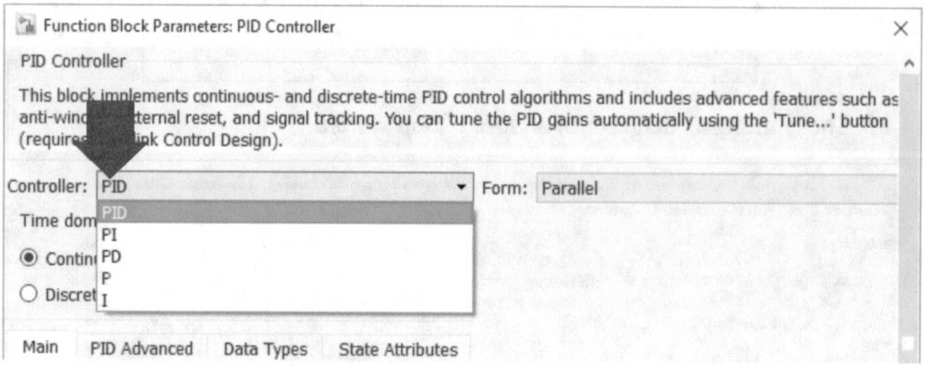

Figure 4.19: Select the desired type of controller from the drop-down list.

Click the Tune... button to start the tuning (Fig. 4.20).

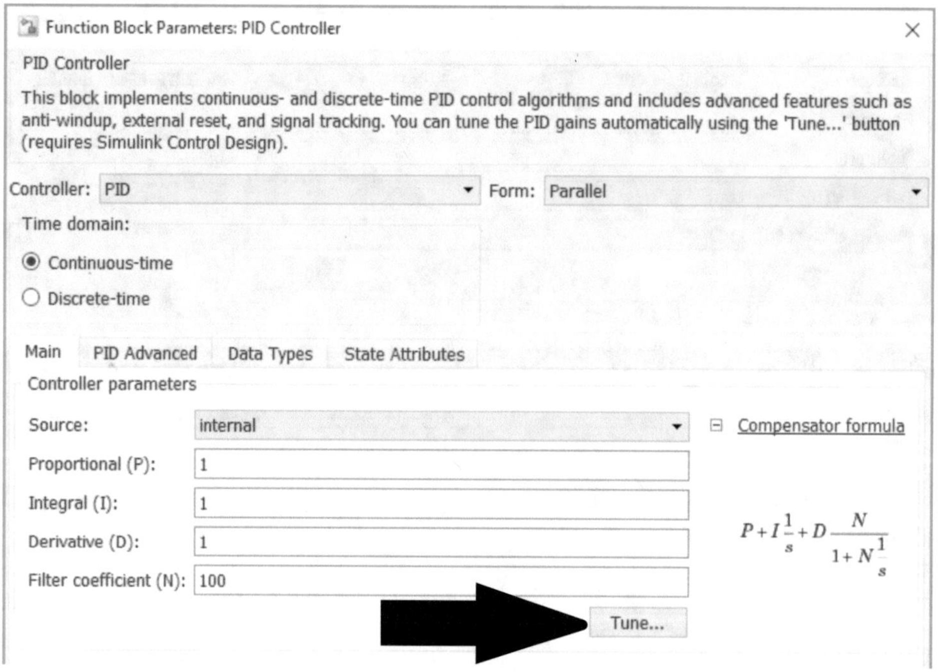

Figure 4.20: Click the Tune…button to start the tuning.

The Simulink® runs the PID Tuner for you. You can tune the block in the same way studied before (Fig. 4.21). Once you obtain the desired output response (with acceptable control effort) you can click the Update Block icon and Simulink® update the PID block with the new coefficients (Fig. 4.22).

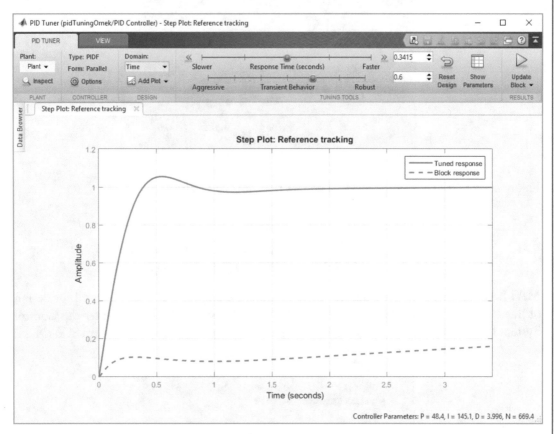

Figure 4.21: Simulink® runs the PID tuner after you clicked the tune ….

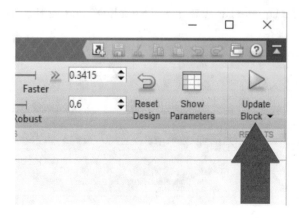

Figure 4.22: Click the update block to update the controller coefficients.

4.4 DESIGN OF LINEAR CONTROLLERS WITH SISOTOOL (CONTROL SYSTEM DESIGNER)

MATLAB® has a application named Control System Designer which can be used for design of linear control systems. You can use the commands shown in Fig. 4.23 to enter the Control System Designer environment. The Control System Designer environment is shown in Fig. 4.24.

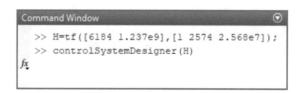

Figure 4.23: Control system designer can be run with the command `controlSystemDesigner`. We want to design a linear controller for H.

You can select the desired structure using the "Edit Architecture" button as shown in Fig. 4.25. Using the icons you can import the transfer functions from the Workspace.

You can use the "Tuning Methods" button (Fig. 4.26) and "Automated Tuning" section to automatically tune the controller (Block C in Fig. 4.25). If you click the "PID Tuning," a window like that shown in Fig. 4.27 will appear.

You can do the tuning either in time domain or frequency domain (Fig. 4.28).

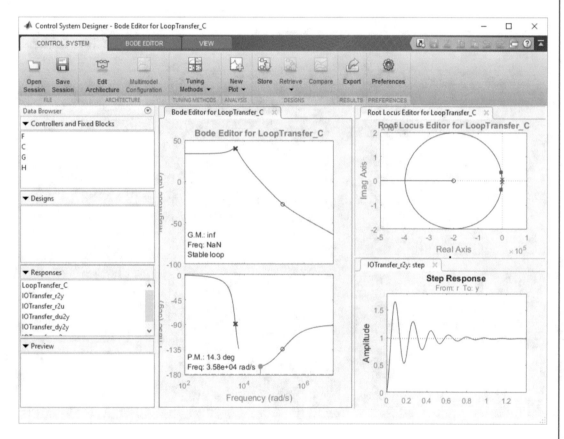

Figure 4.24: "Control system designer" window.

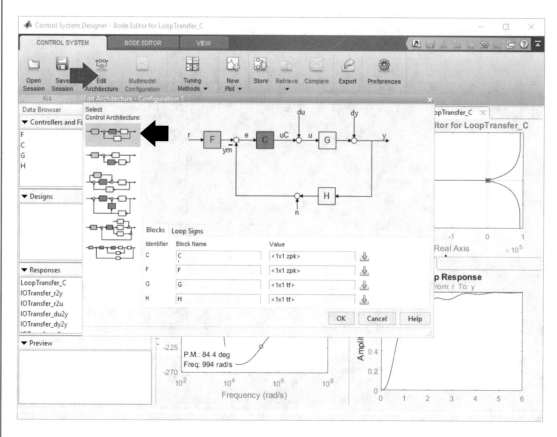

Figure 4.25: Selecting the desired control topology. The one shown with black arrow is the most commonly used topology.

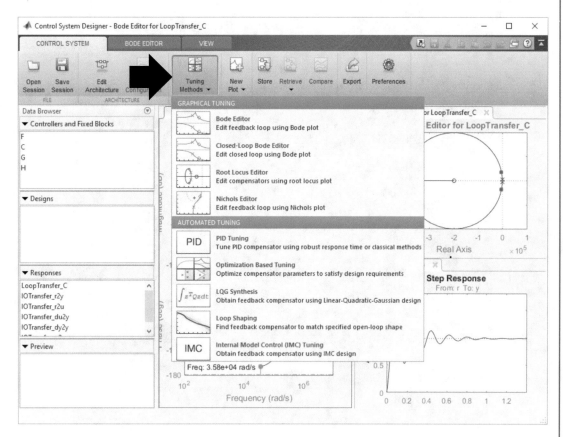

Figure 4.26: "Tuning methods" button.

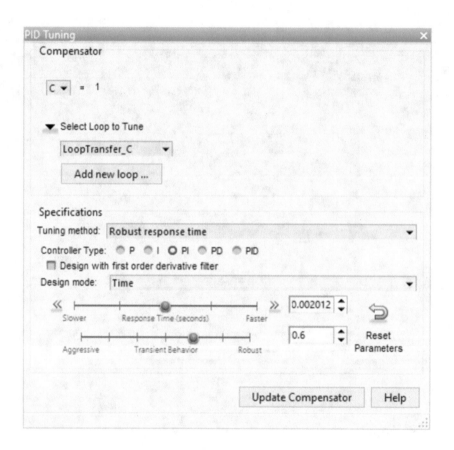

Figure 4.27: **PID** tuning window.

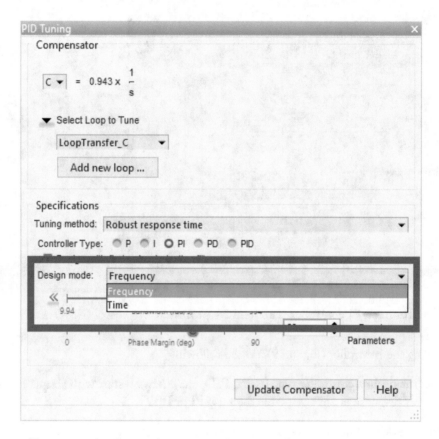

Figure 4.28: Tuning can be done either in time domain or frequency domain.

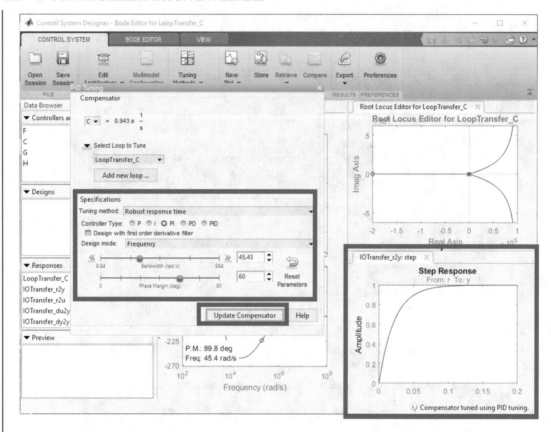

Figure 4.29: Tuning a PI controller is an iterative process.

To tune the controller (Fig. 4.29), do the following.

1. Select the desired controller type (P, PI, PID, etc.). Always start with simple P-type controller. If response is not acceptable then use PI or PID.

2. Select the desired domain (time or frequency).

3. Move the sliders.

4. Click the Update Compensator Button.

5. See the step response and decide whether the response is good or not. If you are not happy, redo the previous steps, i.e., move the sliders again.

You can right-click on the shown step response and use the Characteristics in order to see the characteristics of the response (Fig. 4.30).

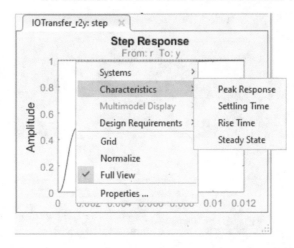

Figure 4.30: Reading the step response characteristics.

You can export the tuned controller to Workspace. To do this, click the Export button (Fig. 4.31).

Figure 4.31: Exporting the controller to workspace.

A window like the one shown in Fig. 4.32 will appear. Check the box to the left of C and click the Export button.

A new variable named C will be added to Workspace after you clicked Export button. It contains the designed controller. You can see the controller equation with the aid of tf command (Fig. 4.33).

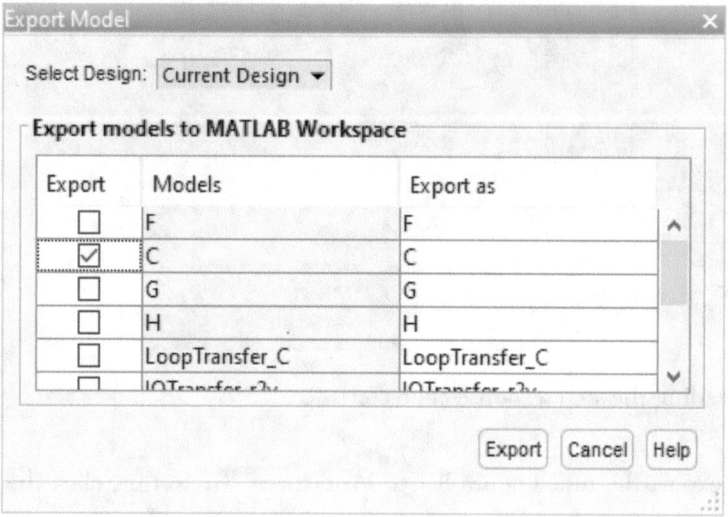

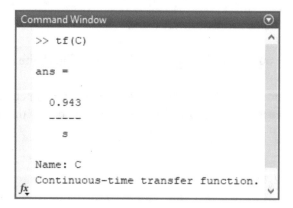

Figure 4.32: **Export model window.**

Figure 4.33: **Equation of designed controller.**

4.5   LOOP SHAPING

You can use design the controller by using the loop shaping method (Fig. 4.34).

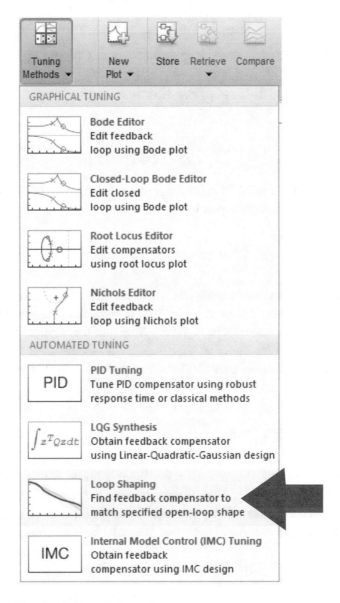

Figure 4.34: Loop Shaping button.

Assume a control structure like that shown in Fig. 4.35. C(s) and P(s) show controller and plant transfer function, respectively.

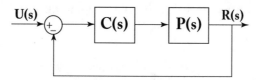

Figure 4.35: Schematic of a simple closed loop control system. C(s) and P(s) show the controller and plant, respectively.

The closed-loop transfer function is:

$$H_{CL}(s) = \frac{R(s)}{U(s)} = \frac{C(s)\,P(s)}{1 + C(s)\,P(s)}.$$

The product of $C(s) \times P(s)$ is called "loop transfer function." Assume that $C(s)\,P(s) = \frac{K}{s}$. In this case the closed-loop transfer function is

$$H_{CL}(s) = \frac{R(s)}{U(s)} = \frac{\frac{K}{s}}{1 + \frac{K}{s}} = \frac{K}{s + K}.$$

If we apply a step function to such a system, it tracks the step function with zero steady-state error within $\frac{5}{K}$ s. For example, if $C(s)\,P(s) = \frac{1000}{s}$, it takes about $\frac{5}{1000} = 5$ ms, to track the step command. The response is shown in Fig. 4.36. Note that the response has no overshoot or oscillatory nature.

As another example, assume that $C(s)\,P(s) = \frac{\omega_n{}^2}{s(s+2\zeta\omega_n)}$. In this case, the closed-loop transfer function is

$$H_{CL}(s) = \frac{R(s)}{U(s)} = \frac{\frac{\omega_n{}^2}{s(s+2\zeta\omega_n)}}{1 + \frac{\omega_n{}^2}{s(s+2\zeta\omega_n)}} = \frac{\omega_n{}^2}{s^2 + 2\zeta\omega_n s + \omega_n{}^2}.$$

Based on the values of ζ and ω_n, the response takes different forms. Figure 4.37, shows the step response for two different values of ζ. As you see, an increase in ζ decrease the oscillatory nature.

Loop shaping takes the desired loop transfer function (i.e., $H_{desired}(s)$) and design the controller block (block C in Fig. 4.35) such that the error between $H_{desired}(s) - C(s)P(s)$ is minimized.

One may ask, $C(s) = \frac{H_{desired}(s)}{P(s)}$ makes the error zero since,

$$H_{desired}(s) - C(s)P(s) = H_{desired}(s) - \frac{H_{desired}(s)}{P(s)}P(s) = 0.$$

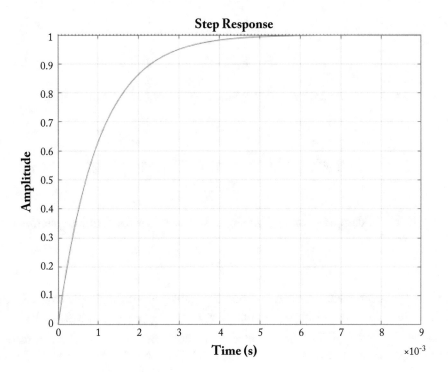

Figure 4.36: Step response of $\frac{1000}{s+1000}$.

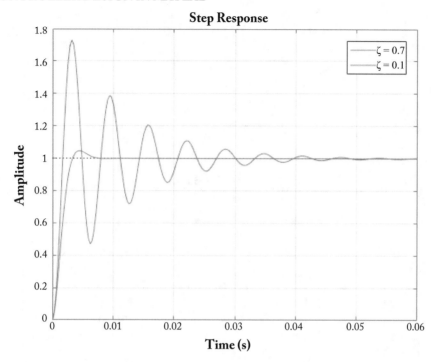

Figure 4.37: Effect of damping ratio (ζ) on step response of a second-order system.

Although the approach seems correct, it is not applicable except for simple cases. Studying an example is quite usefull. For example, assume that $P(s) = \frac{4}{(s+7)(s+8)}$ and $H_{desired}(s) = \frac{10}{s}$. So, $C(s) = \frac{H_{desired}(s)}{P(s)} = \frac{10(s+7)(s+8)}{4s}$. The obtained controller is not proper, so it is not realizable.

The Loop Shaping button in "AUTOMATED TUNING" section of Fig. 4.34, uses optimization techniques to obtain the proper controller C(s). User translates the requirement into the $H_{desired}(s)$ for example if one need a response free of overshoot with settling time less than 5 ms, he/she may use $H_{desired}(s) = C(s)P(s) = \frac{1000}{s}$ is a good option. If an overshoot of less than 5% is acceptable, one may use $H_{desired}(s) = C(s)P(s) = \frac{1000^2}{s(s+2\times0.7\times1000)}$. Normally, $H_{desired}(s)$ is selected among the first- and second-order transfer functions to avoid increase in the order of designed controller.

When you click the Loop Shaping button in Fig. 4.34, the window shown in Fig. 4.38 will appear. You enter the selected $H_{desired}(s)$ into the "Target open-loop shape (LTI)" box using the command "tf." "Frequency range for loop shaping [wmin,wmax]:" box takes the frequency portion where the optimization must be done. Normally, we need the overlap in the low frequency range.

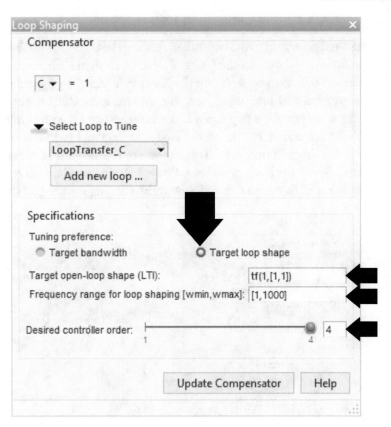

Figure 4.38: Loop Shaping window.

"Desired controller order" slider bar takes the controller order. You start by lower values (i.e., 1). Design process is started by pressing the "Update Compensator" button. When the order of controller is too low, the software shows a warning and asks you to increase the order. In this case you increase the order by one. If the warning comes out again, you increase it again.

4.6   MANUAL CONTROLLER DESIGN

You can add your desired pole/zero to the controller. Adding the pole/zero can be done by right-clicking one of the diagrams shown in the Control System Designer environment (Fig. 4.39). You can add the desired pole/zero by selecting the "Add Pole/Zero" as shown in Fig. 4.40. As shown in Fig. 4.40 you can add integrator, lead, lag, etc. easily by clicking the corresponding choice. If the steady-state error for a step input is not zero, add an integrator to the loop.

You can push the left mouse button on the Bode diagram and (without releasing the left mouse button) move the diagram up/down. This lets you to set the gain of system.

The P.M. and G.M. on the Bode plot shows Phase Margin and Gain Margin, respectively. You can see the equation of designed controller by right-clicking on the plots and select Edit Compensator (Fig. 4.41).

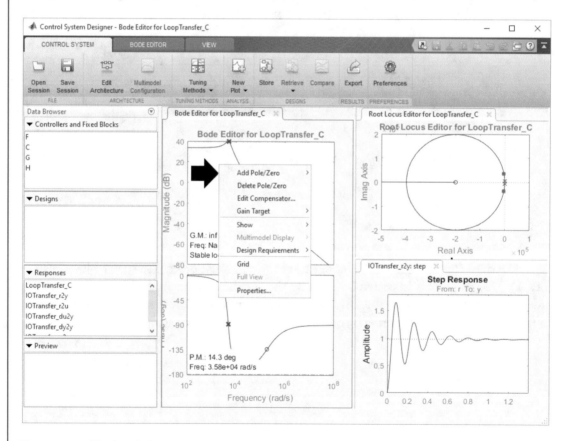

Figure 4.39: Right-clicking on the plot opens the menu.

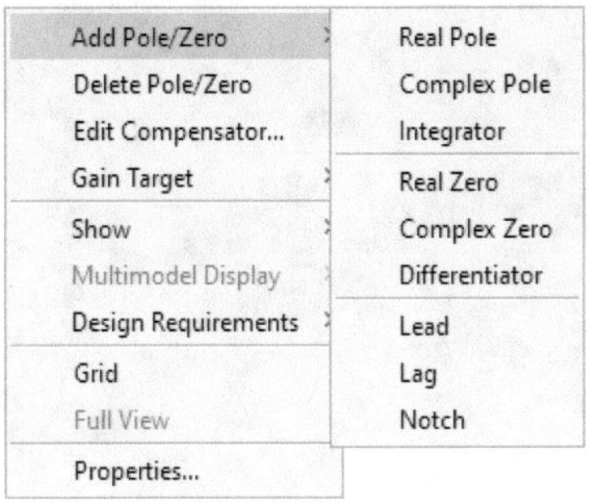

Figure 4.40: "Add pole/zero" menu.

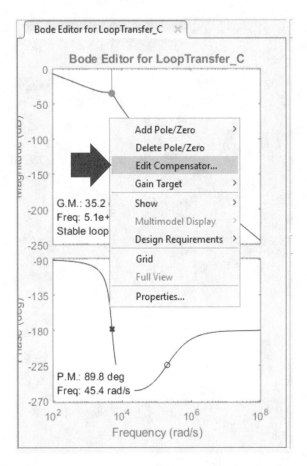

Figure 4.41: Click the edit compensator to see the equation of controller.

CHAPTER 5

Introduction to System Identification Toolbox™

5.1   INTRODUCTION

In the previous chapters we assumed that the dynamical model of plant is known. If you don't know the dynamical model of plant in hand, you can make a model for it based on the input/output data. System identification is a methodology for building mathematical models of dynamic systems using measurements of the system's input and output signals.

This chapter introduces the System Identification Toolbox™ with an example.

5.2   ILLUSTRATIVE EXAMPLE

Consider a system with the following dynamical equation:

$$H(s) = 30\frac{(s+1)(s+2)}{(s+3)(s+4)(s+5)} = \frac{30s^2 + 90s + 60}{s^3 + 12s^2 + 47s + 60}. \tag{5.1}$$

We draw the following simulation diagram (Fig. 5.1).

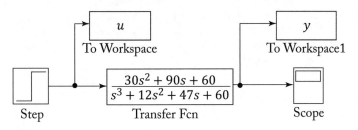

Figure 5.1: Simulink® diagram of the assumed system.

The system is stimulated with a step function. Meanwile the input and output of the system are send to workspace for further process.

The "To Workspace" blocks Save format is set to Structure With Time (Figs. 5.2 and 5.3).

Click the gear icon. Do the setting such as the one shown in Fig. 5.4. This asks the Simulink® to sample the input/output with the rate of 10 ms (Fig. 5.5).

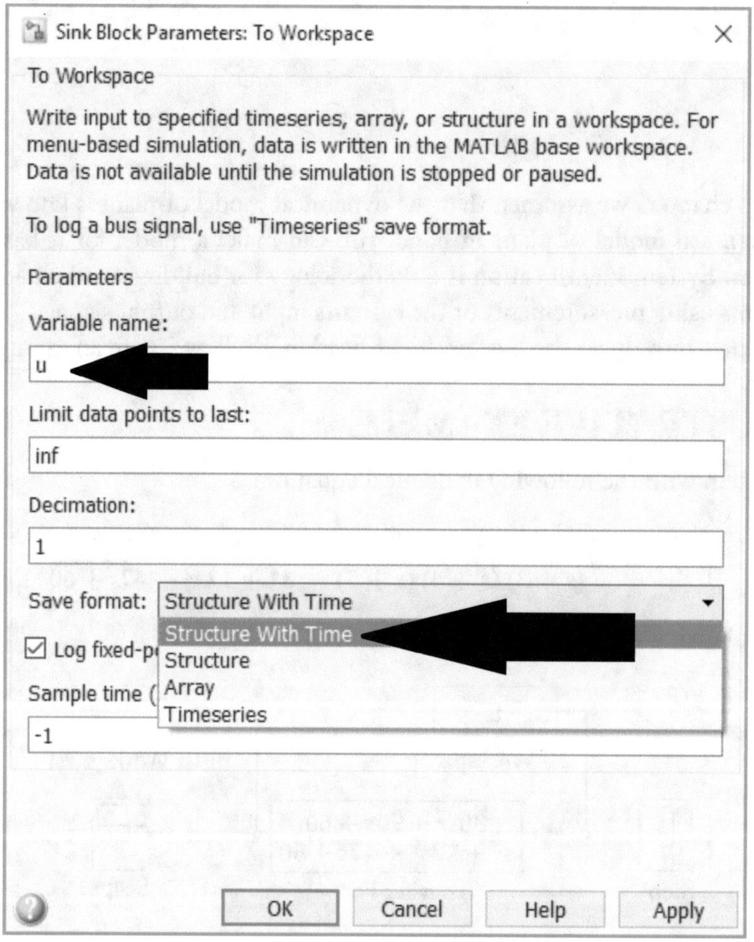

Figure 5.2: **To Workspace** block settings.

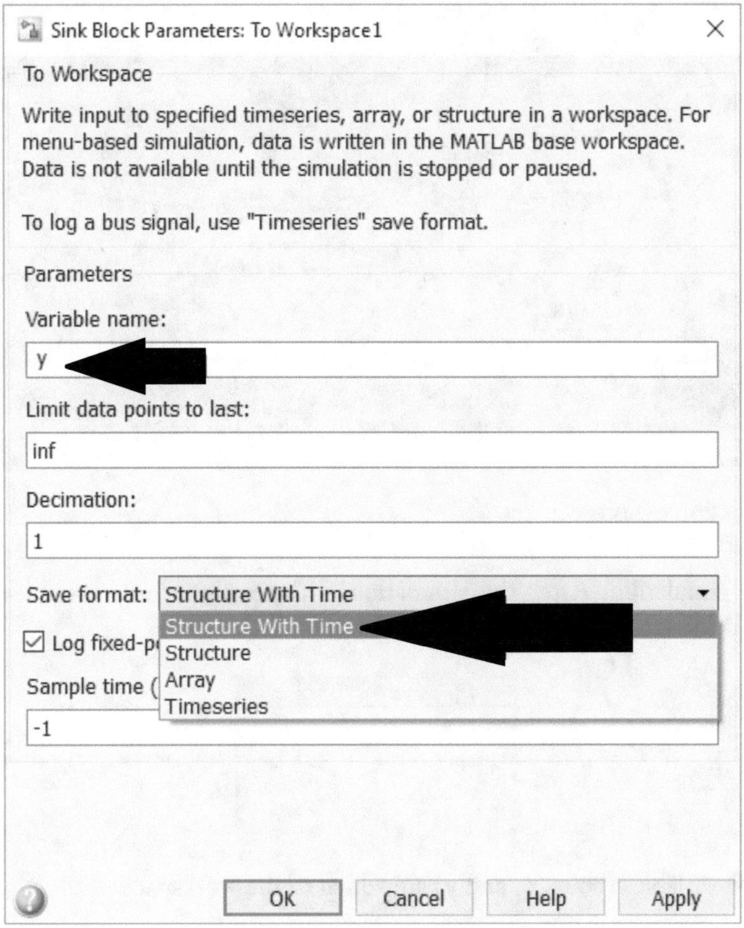

Sink Block Parameters: To Workspace1   ×

To Workspace

Write input to specified timeseries, array, or structure in a workspace. For
menu-based simulation, data is written in the MATLAB base workspace.
Data is not available until the simulation is stopped or paused.

To log a bus signal, use "Timeseries" save format.

Parameters

Variable name:

y

Limit data points to last:

inf

Decimation:

1

Save format: Structure With Time   ▼

              Structure With Time
☑ Log fixed-p Structure
              Array
Sample time (Timeseries

-1

              OK   Cancel   Help   Apply

Figure 5.3: **To Workspace1 block settings.**

Figure 5.4: Model configuration parameter icon.

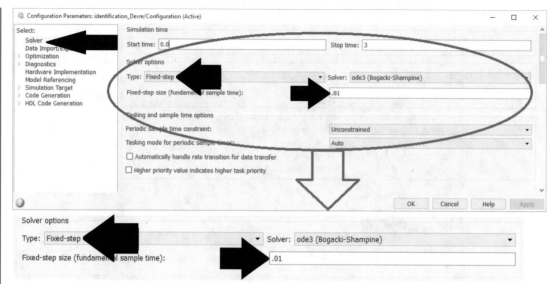

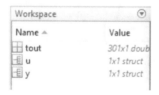

Figure 5.5: Selection of solver.

Run the simulation. After the simulation is done, three new variables are added to MATLAB® Workspace. These are tout, u, and y (Fig. 5.6).

Figure 5.6: New variables (tout, u, and y) are added to the workspace.

u and y are structures. The following commands extract their data.

```
>>U=u.signals.values;
```

```
>>Y=y.signals.values;
```

The following command draws the input and output on the same graph (Fig. 5.7).

```
>>plot(tout,U,tout,Y),
```

```
grid minor,xlabel(`Time(s)'), title(`Input---output of system')
```

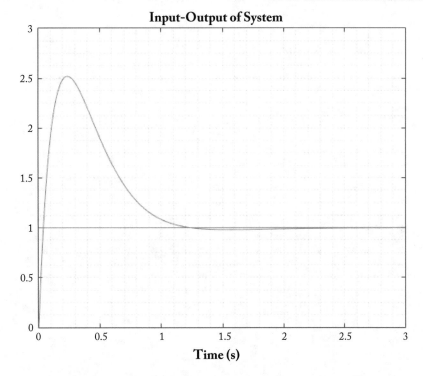

Figure 5.7: Input/output signals of system (Fig. 5.1) are drawn on the same graph.

Run the System Identification Toolbox™ by typing the following command.

```
>>ident
```

The result is shown in Fig. 5.8.

Click the **import data>Time domain data…** to enter the signals into the System Identification Toolbox™ (Fig. 5.9).

After you clicked the **import data>Time domain data…**, the Import Data window will be shown. Fill the boxes as shown in Fig. 5.10 and click the Import and Close buttons, respectively.

Check the Time plot to see the graph of imported data. The graph is shown in Fig. 5.11.

Click the **Estimate> Transfer Function Models …** in order to make a transfer function model for the imported data (Fig. 5.12).

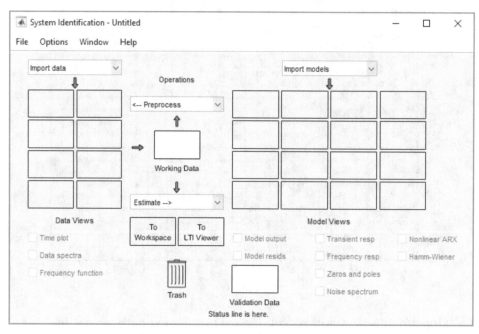

Figure 5.8: System identification Toolbox™.

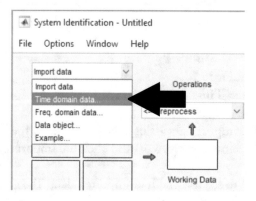

Figure 5.9: Importing the time domain data.

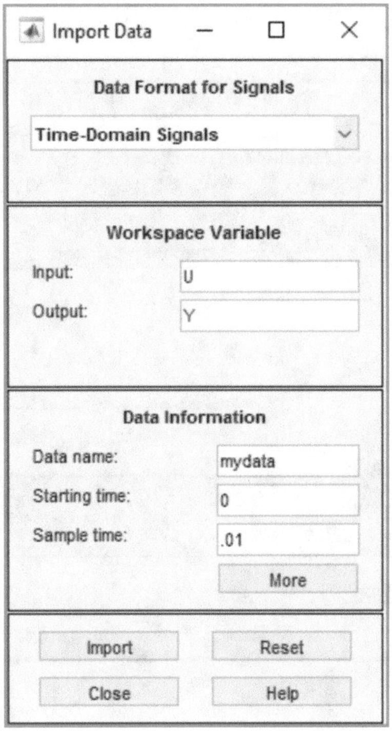

Figure 5.10: Import data window.

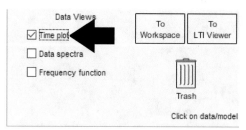

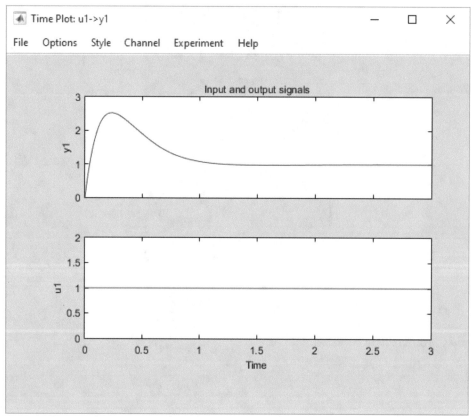

Figure 5.11: Time domain plot of input/output signals.

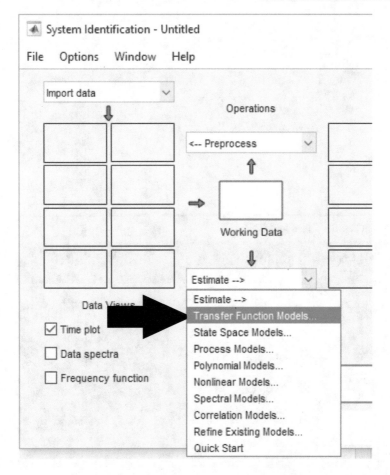

Figure 5.12: Select transfer function models... to estimate a transfer function for the input/output data.

Set the Number of poles and zeros (Fig. 5.13). You must have some insight about your system to set these values, i.e., you know the order of your system. We asked the toolbox to make a third-order model for us. After you enter the number of polse and zeros, click the Estimate button. The toolbox start processing (Fig. 5.14).

After calculations finished, you can see the summary of what has been done in the Result section of the window (Figs. 5.14 and 5.15). Fit to estimation data must be a larg number (i.e., 80%). It can be an indicator of how reliable our model is. Since we have no noise and we set the number of poles and zeros, it reaches 100% in this example. However, in a real-world application, i.e., when the data is collected with sensors, there are some noise and reaching 100% seems difficult.

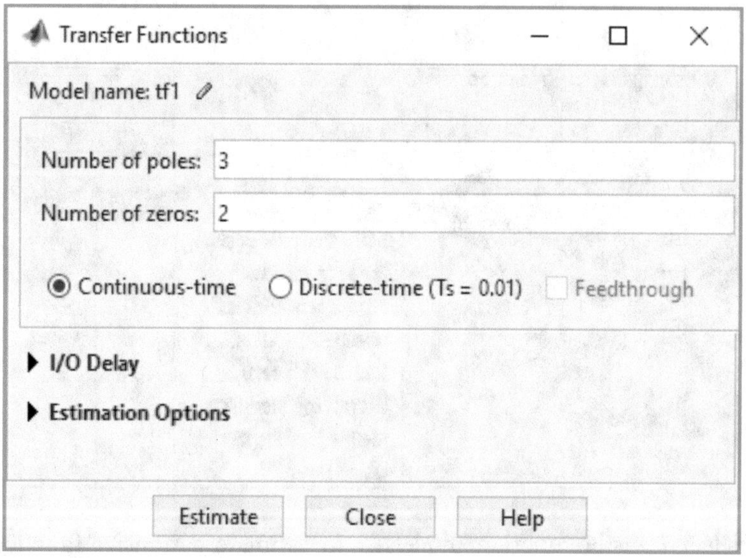

Figure 5.13: Defining the order of model, i.e., how many poles/zeros the model have.

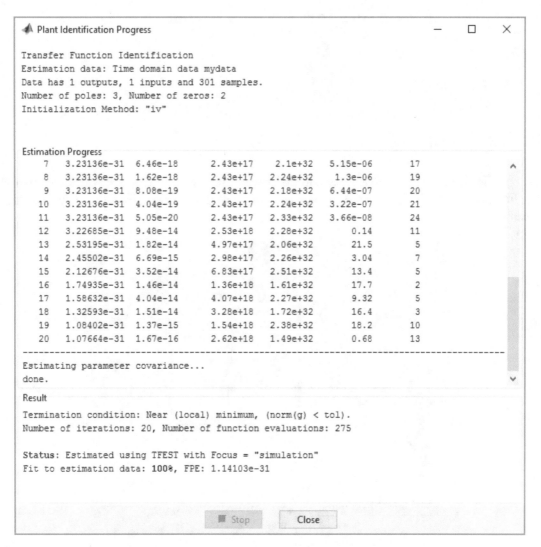

Figure 5.14: Plant identification progress window shows the iterations.

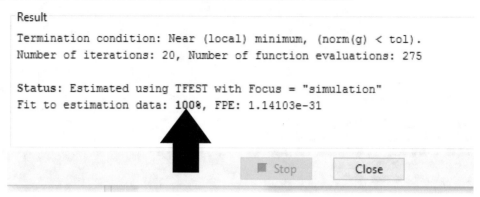

Figure 5.15: Results of calculations.

Click Close to return to main window of toolbox. If you want to send the calculated transfer function to MATLAB® workspace, you can drag the tf1 and release it on the To Workspace button (Fig. 5.16).

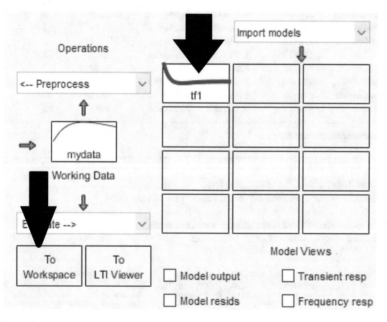

Figure 5.16: Transfering the obtained model to Workspace.

You can see a variable named tf1 is added to the Workspace (Fig. 5.17).
Type tf1 at the Workspace to see the obtained transfer function equation (Fig. 5.18).

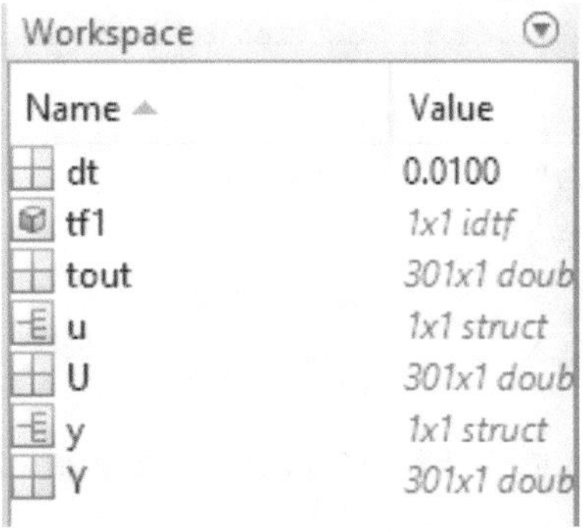

Figure 5.17: New variable named `tf1` is added to the workspace.

```
tf1 =

  From input "u1" to output "y1":
     30 s^2 + 90 s + 60
  -------------------------
  s^3 + 12 s^2 + 47 s + 60

Name: tf1
Continuous-time identified transfer function.

Parameterization:
   Number of poles: 3   Number of zeros: 2
   Number of free coefficients: 6
   Use "tfdata", "getpvec", "getcov" for parameters and their uncertainties.

Status:
 Estimated using TFEST on time domain data "mydata".
 Fit to estimation data: 100% (simulation focus)
 FPE: 1.141e-31, MSE: 9.509e-31
```

Figure 5.18: The equation of estimated model.

If you change the number of poles to 2 and number of zeros to 1, the Fit to estimation decrease a little bit (Fig. 5.19).

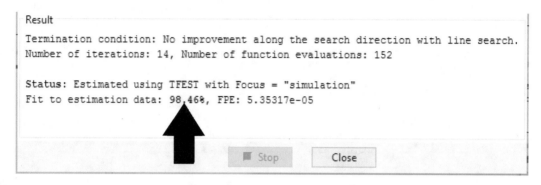

Figure 5.19: Estimating a second-order model to the imported data. In this case the Fit to estimation decreased slightly.

In this case the toolbox gives the following model:

$$H_{2pole,1zero}(s) = \frac{29.4s + 26.24}{s^2 + 9.438s + 26.51}.$$

(5.2)

The following code compares Bode plot of this transfer function with the original transfer function. The result is shown in Fig. 5.20. As shown, the two transfer functions overlap.

```
>> H=tf([30 90 60],[1 12 47 60]);      %original system

>> I=tf([29.4 26.24],[1 9.438 26.51]); %H2pole,1zero

>> bode(H,I), grid on
```

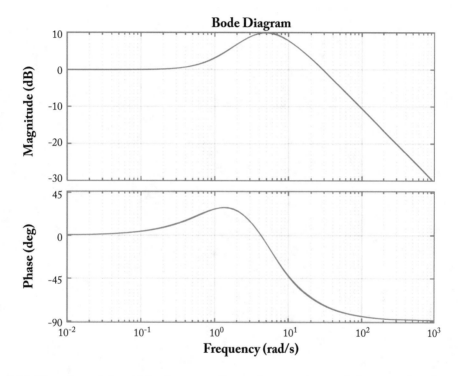

Figure 5.20: Drawing of the original systems frequency response and the second-order estimate on the same graph.

References

[1] Katsuhiko Ogata, *Modern Control Engineering*, Pearson, 2009. DOI: 10.1115/1.3426465.

[2] Benjamin C. Kuo and Farid Golnaraghi, *Automatic Control Systems*, Wiley, 2002.

[3] Gene F. Franklin, J. Da Powell, and Abbas Emami Naeini, *Feedback Control of Dynamic Systems*, Pearson, 2014.

[4] Chi-Tsong Chen, *Linear System Theory and Design*, Oxford University, 2012.

[5] Katsuhiko Ogata, *Matlab for Control Engineers*, Pearson, 2007.

[6] *Simscape User's Guide*, MathWorks, 2015.

[7] Lennart Lejung, *System Identification: Theory for the User*, Prentice Hall, 1999.

[8] Johan Schoukens, Rik Pintelon, and Yves Rolain, *Mastering System Identification in 100 Exercises*, Wiley, IEEE Press, 2012. DOI: 10.1002/9781118218532.

[9] Katsuhiko Ogata, *Matlab for Control Engineers*, Pearson, 2007.

[10] Brian Hahn and Daniel T. Valentine, *Essential Matlab for Engineers and Scientists*, Academic Press, 2016.

[11] Farzin Asadi, *Computer Techniques for Dynamic Modeling for DC-DC Power Converters*, Synthesis Lectures on Power Electronics, Morgan & Claypool Publishers, 2018.

[12] Farzin Asadi, *Robust Control of DC-DC Converters: The Kharitonov's Theorem Approach with MATLAB® Codes*, Synthesis Lectures on Power Electronics, Morgan & Claypool Publishers, 2018.

[13] Farzin Asadi and Key Eguchi, *Dynamics and Control of DC-DC Converters*, Synthesis Lectures on Power Electronics, Morgan & Claypool Publishers, 2018.

[14] Farzin Asadi, Sawai Pongswadt, Kei Eguchi, and Ngo Lam Trung, *Modeling Uncertainties in DC-DC Converters with MATLAB® and PLECS®*, Synthesis Lectures on Power Electronics, Morgan & Claypool Publishers, 2018.

Authors' Biographies

FARZIN ASADI

Farzin Asadi is with the Department of Mechatronics Engineering at the Kocaeli University, Kocaeli, Turkey. Farzin has published 30 international papers and 10 books. He is on the editorial board of 6 scientific journals as well. His research interests include switching converters, control theory, robust control of power electronics converters, and robotics.

ROBERT E. BOLANOS

Robert E. Bolanos received a B.S. in Electrical Engineering from the University of Texas at San Antonio, and an M.S. in Electrical Engineering from the University of Idaho. Currently, he is a Principal Engineer at Southwest Research Institute, Space Science and Engineering Division 15. Mr. Bolanos designs space-rated switch power supplies, RF oscillators, and high voltage pulsers. He has worked on both NASA and ESA space projects such as IBEX, JUICE, BepiColombo, and EUROPA (MASPEX).

JORGE RODRÍGUEZ

Jorge Rodríguez received a B.S. in Electronics Engineering and a Master's in Electronics Engineering, both at Carlos III University of Madrid (Spain). Currently, he is working as a R&D hardware engineer at Power Smart Control focusing on SoC devices, developing HIL systems, and as a support engineer for PSIM and SmartCtrl software.

Printed in the United States
by Baker & Taylor Publisher Services